Table of Contents

Heart Notes

An Informational Guide to Help You to Better Understand Cardiac Diseases, Conditions and the Procedures Used in Their Diagnosis and Treatment

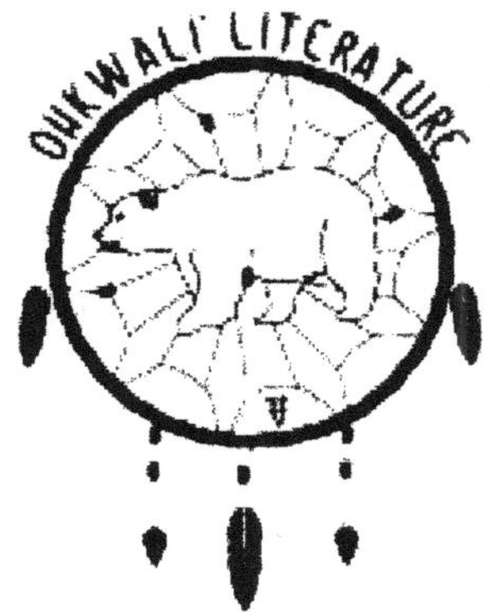

Printed and Bound in the United States of America

Heart Notes

Published by:
Dennis Fellows
Ohkwali' Literature
ohkwaliliterature@verizon.net
ISBN # 978-0-615-25768-6
First Printing 2008

Warning & Disclaimer:

This book, "Heart Notes," was written to help individuals acquire a better understanding of cardiac conditions, diseases and rhythms. It has been designed to serve as a reference tool only and the information provided is not to be considered as absolute or definite. The author has conducted extensive book research and personal consultation with nursing and other medical personnel in an effort to provide complete and accurate information; however there may be undetected errors or mistakes. All judgments regarding the diagnosis and treatment of cardiac related conditions and abnormalities should be made by individuals that have had many years of formal education and training in the field related to the diagnosis and treatment of the heart such as Cardiology Specialists, Electrophysiologists and Medical Doctors. This author will not be held responsible or liable to any individual or entity for damage or harm caused by, or allegedly caused by this book.

Ohkwali' Literature 2008

Introduction:

Today's well educated healthcare personnel and technological advances in diagnostic testing have made the treatment of cardiac diseases and conditions precise and quick. However, there is confusion that is created by the lack of knowledge as to why certain diagnostic tests are performed and in the comprehension of the results obtained. This book was written to assist both medical personnel and laypersons in understanding the how's and why's of the tests that are being utilized and to give a basic description of the diagnosing procedures, diseases and conditions. As with everything, to fully understand any one concept concerning cardiac issues it is helpful to be able to follow the footprints that caused them to reach their present state. This book attempts to do just that by providing a basic explanation of both the intrinsic and non-intrinsic factors that may have contributed to the onset or the progression of various conditions and diseases. As each question is answered and different abnormalities are explained and understood, your understanding surrounding personal or patient issues will increase. The new knowledge will help you in communicating with your healthcare provider and enable you to formulate the proper inquiries to reach the level of understanding that you desire. It is my hope that when you read this book it will give you a better understanding of cardiac abnormalities and their possible causes, help to alleviate fear and confusion of unknown procedures and that it will encourage and help you to establish behavioral changes that could help to prevent self induced illness that improper habits would otherwise have caused to occur in the future.

Sincerely,

Dennis Fellows

Understanding How it Works

The heart is a hollow muscle that serves as a pump to supply oxygen-rich blood to the lungs and body, providing it with what it needs to function. It is made-up of four different chambers, two smaller upper chambers called the right and left atria, and two larger lower chambers that are called the right and left ventricles. The chamber thickness is three times thicker in the left ventricle compared to the right ventricle due to the increased pressure demand on the left side to produce a greater catalyst to perfuse blood throughout the body. The two sides of the heart are divided by a thick muscle that is called the septum. It has a regulated blood flow control system that consists of four valves to control the flow of blood through the heart and to the lungs and body. The tricuspid valve is located between the right atria and right ventricle. This valve has three flaps and is responsible for controlling the flow of blood transfer between these two chambers. The left atria to left ventricle blood flow is managed by the mitral valve, also called the bicuspid valve. This valve possesses two flaps. As the blood exits the right and left ventricles to be sent to the body and lungs, it passes through two valves that have two flaps each known as semi-lunar valves. These valves are designed to provide a positive seal under the pressure within the chambers during the depolarization phase of the cardiac cycle. There are nodes that possess specialized cells that create and transmit electrical impulses through an impulse pathway within the myocardium to induce reflex actions, these are called the sinoatrial node and the atrioventricular node. The sinoatrial node is located in the right atria and is the origin of the electric impulse. From this point it is transmitted to the atria. The atrioventricular node regulates the transmission of the electrical impulse from the atria to the ventricles, and is located between the upper chambers of the heart or the atria, and the lower chambers of the heart or ventricles. The electric impulses transmitted through this pathway cause the muscular structure of the heart to contract or squeeze in a rhythmatic fashion. In the cardiac cycle the right atria receives a supply of waste blood from the body and the left atria has received a supply of freshly re-oxygenated blood from the lungs. The electrical impulse is delivered by the sinoatrial node to the atria and travels through an intra-atrial pathway where it depolarizes the right and left atrial chambers simultaneously. At this point the atrial chambers contract and squeeze their fill of blood. The left atria transfers its fill of blood through the mitral valve to the left ventricle, and the right atria sends its fill of blood to the right ventricle through the tricuspid valve. With the ventricular chambers full of blood the electrical impulse then travels along an intra-nodal pathway through the atrioventricular node and the bundle of his, down the right and left bundle branches to the purkinje fibers. At this point the impulse depolarizes the right and left ventricles forcing them to evacuate their fill of blood. The blood from the left ventricle is perfused through a semilunar valve to the aorta and then to the body and its organs. The right chamber sends its blood through a semilunar valve to the pulmonary artery and then to the lungs for re-oxygenization. This complete process from the initiation of the impulse at the site of the sinoatrial node to the completion of the ventricular repolarization is referred to as the cardiac cycle.

The Cardiac Cycle

The cardiac cycle is a culmination of organized electrical impulses delivered by the hearts sinoatrial (SA) node that travel along a prescribed route to induce muscular contractions. This action facilitates the emptying and refilling of the atrial and ventricular chambers to provide a constant supply of oxygen-rich blood to the body and lungs. This diagram is helpful in describing where the impulse is in the cardiac cycle and what is occurring with each of the different wave forms observed on a rhythm strip tracing.

* The initial impulse is sent by the Sinoatrial node to the atria by way of an interatrial pathway that traverses from the right to the left atria to depolarize both of the upper chambers of the heart to begin the cardiac cycle. This action causes a muscular reflex within the atria which causes them to contract and move their fill of blood to the lower chambers called the ventricles.

* The electrical impulse then proceeds through a prescribed intranodal pathway to the AV junction where it passes through the atrioventricular node, bundle of His, both the right and left bundle branches to the purkinje fiber network where it stimulates the ventricles to contract and squeeze their fill of blood to the lungs and body. The atria also repolarize at this time to rest and receive a new fill of blood.

* The ventricles then rest and the cardiac cells recharge for the next cardiac cycle. At this point in the chambers also receive a fresh supply of blood to be delivered to the body and lungs to be delivered during the next cardiac cycle. lungs in the next cycle.

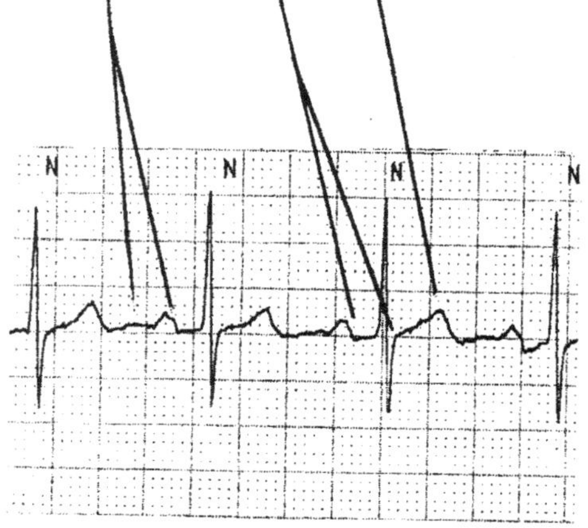

These cycles in the normal heart will occur at an average rate of 70 times per minute (between 60 and 100) cycles per minute, 100,800 times per day, pumping 5 quarts of blood being through its chambers every 60 seconds.

The Conduction System

The conduction system of the heart begins with an electrical impulse that originates and is delivered from the sinoatrial node. This node is made up of a cluster of specialized cells/tissue and is located in the upper section of the right atria. As with the other cells throughout the heart these cells possess automaticity which gives them the ability to generate their own electrical impulses to induce myocardial muscle depolarization without the aide of outside stimulation. However, the sinoatrial node generates electrical impulses and discharges them at a rate that exceeds the discharge rates of the other cardiac cells located in the AV junction and purkinje fibers within the ventricles. This causes the impulse to be transmitted through the myocardium at a point when the other cells are completing their repolarization. It is therefore given the term "heart's natural pacemaker." The conduction system is in essence a rhythmic series of regulated impulses created and delivered by the sinoatrial node that travel through the myocardium inducing muscular responses in the form of contractions. Under normal conditions the impulse is generated in the sinoatrial node and is transmitted through an inter-atrial pathway to stimulate both the right and left atria which causes them to contract in synchronization and push their fill of blood through the mitral and tricuspid valves into the ventricles. This impulse is then transmitted along an intra-nodal pathway to the atrioventricular node (AV node) where there is a slight delay to allow chamber to chamber filling time from the atria to the ventricles, through the bundle of his which is a nerve pathway located between the right and left ventricles, down the right and left bundle branches to the purkinje fibers where it stimulates the ventricles to depolarize and push their fill of blood through the pulmonary and aortic valves (semi-lunar) to the lungs and body. Following depolarization the cardiac cells recharge, a phase that is called the repolarization or refractory period. The refractory period has three variations that are referred to as the absolute, relative, and supernormal. Each of these stages within the repolarization process presents cardiac cells at different thresholds of vulnerability. The absolute refractory period is timed from the beginning of the ventricular depolarization to the center of the T-wave or midway point of the ventricular repolarization process. This phase presents cells in a state that will not conduct or respond to electrical stimulation. The next stage is the relative refractory period which is timed from the center of the T-wave or midway through the repolarization process to the point where it slopes down and merges with the isoelectric baseline which presents cardiac cells that have recharged to a degree that an introduction of electrical stimulation with enough intensity can induce a ventricular response. The last level is the supernormal refractory period and begins at the end of the T-wave. The cells in this refractory period have repolarized to a threshold where normal electrical stimulation will precipitate a ventricular response. It is during the relative and supernormal refractory periods that triggered automaticity from early electrical impulses induces arrhythmias such as Torsades de Pointes. This process, from the initial electrical impulse created at the site of sinoatrial node to the completion of the ventricular repolarization is referred to as the cardiac cycle. The following descriptions are the identifiable characteristics associated with the hearts conduction cycle as seen on a rhythm strip tracing.

EKG Characteristics

P-Wave Indicates that an atrial depolarization has occurred. In normal sinus rhythm the P-wave is upright and rounded. Important factors in evaluating the P-wave are its configuration, deflection and location. P-waves that are wide and possess a notch or present peaked may indicate diseases such as congestive heart failure, chronic obstructive pulmonary disease, or abnormalities or defects related the heart valves. The junctional rhythm possesses inverted P-waves that can be located before, within, or after the QRS; with its position in reference to the QRS within the cycle indicative as to the order of the atrial and ventricular depolarization. P-waves that present with three different morphologies that originate from different focal points within the myocardium indicate the presence of an arrhythmia that is referred to as wandering atrial pacemaker.

PR Interval Represents the time duration it takes for the electrical impulse to initiate a ventricular depolarization as measured from the time that the atrial depolarization began. It is measured from the beginning of the P-wave to the beginning of the QRS with the normal sinus rhythm measurement being 0.12-0.20 one hundredths of a second. A measurement greater than 0.20 indicates that there was a delay in impulse transmission time from the atria to the ventricles and is referred to as a 1st degree atrioventricular block or 1st degree AV block. A measurement less than 0.12 of the PR interval accompanied by an inverted P-wave is indicative of a retrograde impulse that originated from the AV junction. PR intervals that are abnormally short in time duration may also indicate that impulses may have utilized an accessory pathway from the atria to the ventricles outside of the AV node, such as those seen in the anomaly called Wolff Parkinson White Syndrome or WPW. In this myocardial abnormality the impulses take an alternate unregulated pathway that is a direct link from the atria to the ventricles known as the Bundle of Kent.

QRS This pattern on the EKG consists of a Q-wave, R-wave and S-wave. These waves indicate that ventricular depolarization has taken place. The normal sinus rhythm time duration measurement of the QRS is 0.06-0.10 one hundredths of a second. A time measure greater than 0.12 one hundredths of a second indicates that there was a delay in the impulse conduction time from the AV junction to the purkinje fibers and is referred to as a bundle branch block. When evaluating, focus on the configuration and time duration.

ST Segment This represents that the ventricular depolarization has ended and the repolarization phase of the cardiac cycle is beginning. This segment is the space located between the S-wave and the beginning of the T-wave, with the actual point where the transition from the ventricular depolarization and repolarization begins referred to as the J-Point. An observation of elevation of this segment may indicate transmural myocardial injury, with segment depression indicative of subendocardial injury.

T-Wave This represents the ventricular repolarization period. They should appear upright, smooth and rounded. When evaluating the T-wave you should focus on the T-wave's configuration and deflection. Inverted T-waves are seen in situations such as transmural ischemia while peaked T-waves may indicate hyperkalemia.

QT Interval This is the time duration of the ventricular depolarization and repolarization. If this interval becomes prolonged as it does in the anomaly called Long QT Syndrome, there is an opportunity for an early ectopic beat to enter the vulnerable ventricular cells during the relative or supernormal refractory periods of the repolarization phase. This event can lead to the stimulation of a life threatening arrhythmia such as Torsades de Pointes through an event called triggered automaticity.

Aneurysm

An aneurysm is a protrusion in a compromised wall section of a blood vessel that has been weakened due to plaque build-up, high blood pressure, or a congenital abnormality that has been present since birth. These blood filled bulges present differently and are classified according to their location and appearance. One example would be the fusiform aneurysm in which there is an egg shaped swelling that encompasses the entire diameter of a section of the blood vessel. The fusiform aneurysm is not bubble shaped because the internal vessel pressure pushes out against all three layers of the vessel wall uniformly with the outer protective layer weakened but intact. The saccular aneurysm occurs when the inner layers of the vessel squeeze through a tear in the outer layer of the vessel. This gives the appearance of a balloon on the side of the vessel and is considered to be the definition of the real aneurysm. Aside from their appearance aneurysms can develop in different locations of the cardiovascular system. The cerebral aneurysm forms in the arteries located at the base of the brain in a spot referred to as the "Circle of Willis." Aortic aneurysms can be found in two different locations within the wall of the aorta which is the main artery that receives freshly re-oxygenated blood from the left ventricle to be delivered to the body. They are referred to as the "thoracic aortic aneurysm" or "TAA" which is located in the chest area and the "abdominal aortic aneurysm" or "AAA" that is found in the abdominal section of the aorta. Cirsoid aneurysms are enlarged blood vessels that appear on the scalp that can eventually evolve into a tumor. Aneurysms are precipitated by conditions such as high blood pressure where there is a constant uncontrolled pressure created within the arterial lumens that puts stress that pushes out against the inner lining of the arterial vessel. This will eventually cause weakening or ruptures in the vessel walls that will allow the inner linings to balloon through the outer lining such as that seen in the saccular aneurysm, or the bulging of the entire circumference of the vessel as seen in the fisiform aneurysm. Another condition is a disease referred to as coronary artery disease or "artherosclerosis." This disease is the accumulation of fatty deposits in the lumens of the blood vessels that occurs over a period of years that can eventually cause them to weaken and rupture. Unhealthy lifestyle habits such as cigarette smoking, excessive alcohol consumption, or poor dietary practices that can lead to obesity are contributory factors. Aneurysms can be treated using a couple of different methods that range from the implementation of lifestyle changes to control hypertension and diabetes, to the employment of both invasive and minimally invasive surgical procedures. One of the methods is referred to as "surgical clipping." This procedure was first performed at Johns Hopkins Hospital in the year 1937 by Dr. Walter Dandy. This procedure required a craniotomy to present the aneurysm at which point a clip is positioned at the base of the aneurysm which eliminates it from the cardiovascular equation. Another less invasive procedure is called "endovascular coiling" or "coil embolization." This is performed with the use of fluoroscopic imaging, a catheter tube and a platinum coiling. The catheter is fed through the femoral artery at a site located in the groin to the cerebral aneurysm where platinum coil is deployed that will conform to and fill the space within the aneurysm. This blocks the blood flow to the aneurysm and decreases the probability of rupture.

This procedure has been utilized on the aneurysms of greater than 125,000 patients where surgical intervention was not desirable or unattainable. This method is less time consuming, less invasive and requires a shorter recovery time. But this method does present with drawbacks compared to the surgical clipping method. Studies have shown that of the endovascular coiling procedures done that there was a 28% to 33% incidence where the aneurysms recurred within a year. This necessitates the need for the implementation of constant monitoring with tests such as the MRI and MRA to evaluate for possible recurrences. Aneurysms that occur in the peripheral vessels or the aorta are treated surgically using a couple of different methods. One procedure employs the use of a plastic tube to replace the section of the vessel containing the aneurysm. The other method is called an endoaneurysmorrhaphy and involved the opening of the aneurysm sack then folding and suturing the walls to restore vessel integrity. Aneurysms can cause mild symptoms such as back pain if there are any symptoms that are present at all and if ruptured can cause massive internal bleeding; leaving only minutes for treatment. It is estimated that individuals who experience a ruptured aneurysm have a twenty percent chance of survival. It is also documented that men are more prone to the occurrence of aneurysms than women due to the fact that estrogen helps women to absorb copper into their system more readily than men. Decreased amounts of copper in the system have been shown to predispose individuals to the possibility of suffering an aneurysm.

Did You Know?

Coumadin (Warfarin) Anticoagulant

Coumadin is classified into a category of anticoagulants or "blood thinners," but is in actuality not a blood thinner. Coumadin is used as an anticoagulant in situations where there is the potential for the formation of clots present. This drug works as a clot prevention option that represses the synthesis of blood clotting factors that normally combine when there is a break in a blood vessel to stop the bleeding. Although coumadin is successful in helping to prevent clots it does not actually thin the blood and cannot dissolve a clot that is already formed. This drug is administered in situations involving arrhythmias such as atrial fibrillation where there is a predisposition of coagulation and clot formation due to blood pooling caused by decreased circulation and in the prevention and management of DVT or deep vein thrombosis where there is the possibility of clots forming within the deep veins of the legs that can produce pulmonary emboli. The major side effect of this medication is uncontrolled bleeding due to the induced inhibition of clotting factors.

Ambulatory Defibrillator Vest

There is now a vest that is designed to perform as both a data recorder of rhythm and arrhythmia information as well as an external defibrillator. This device is worn like a vest that incorporates a non-invasive portable defibrillator that continuously receives the electrical signals that travel through the heart through wires connected to electrodes that are attached to the torso. When arrhythmias arise that require a corrective electrical shock, the wearer is alerted and given the option of overriding the shock. If the shock is not aborted, the device will automatically deliver the shock up to 150 joules to terminate the arrhythmia. This vest provides a non-invasive, surgery free defibrillator option that due to its rapid response time (under 1 minute), presents with a 75% success rate in preventing sudden cardiac death. This vest is not to be confused with the Holter Monitor that is designed as a diagnostic tool for the recording of myocardial electrical activity only.

<u>Antidysrhythmic Drug Classification Chart</u>

Class I	**Class IA**	**Class IB**	**Class IC**
Moricizine	Disopyramide	Lidocaine	Flecainide
	Procainamide	Mexiletine	Indecainide
	Quinidine	Phenytoin	Propafenone
		Tocainide	

Class II
Acebuterol
Esmolol
Propranolol
Sotalol

Class III
Amiodarone
Bretylium
Ibutilide

Class IV
Verapamil

Other medications utilized in cardiac treatment
Drug

Atropine

Angina

The word angina describes chest discomfort that can manifest itself in different forms. It can present as pain that may radiate to the right arm, shoulders and jaw, or it may be a pressure or squeezing sensation localized in the chest area. Angina is caused by an insufficient amount of oxygen-rich blood that is supplied to the heart muscle that is commonly induced by a condition known as "Coronary Artery Disease." This disease is the accumulation of fatty deposits or plaque build-up within the lumens of the coronary arteries that compromises the amount of oxygen-rich blood that feeds the heart. This condition leads to oxygen starvation and eventually cell death (necrosis) to the area of the heart muscle fed by the compromised artery. There are three different types of angina, they are:

Stable Angina – this is chest discomfort is most noticeable when an individual is under stress such as during exercise or performing tasks that require exertion. This form of angina is brought on by the hearts need for an increased supply of oxygen-rich blood that is not being supplied due to the diminished blood flow. These episodes are normally temporary in nature, lasting for a short time.

Unstable Angina – this is the more dangerous of the two types of angina. This form of angina happen without warning in situations where the individual is at rest and there is no exertion or physical stress occurring. Episodes of unstable angina usually last longer than those created by stable angina and send a more urgent signal of serious impending cardiac problems such as a potential heart attack.

Prinzmetal Angina – this is also called "Variant Angina" and is considered a form of unstable angina that frequently occurs during rest and between the hours of midnight and eight in the morning. This form of angina possesses the capability to lead to the onset of dangerous arrhythmias such as ventricular fibrillation and ventricular tachycardia.

Informational Facts:

* angina is also referred to as "angina pectoris" which means "a strangling sensation in the chest."
* there are an estimated 6.8 million cases of angina reported every year. Of these, 4.2 million are women and 400,000 are reports of new cases of stable angina.
* it is estimated that two-thirds of the individuals afflicted with prinzmetal angina have a critical blockage in at least one of their major vessels that feed the heart.

There are a variety of different reasons that the heart can become stressed to a point where there is a need for more oxygen. It is during these episodes when the need is not met that angina presents. Some of these would include emotional stress created by frustration or anger, consuming a heavy meal, cigarette smoking, exercise and sudden changes in altitude or temperature.

Treating angina depends upon its severity. Some methods that may be utilized include lifestyle changes such as controlling diabetes, weight, and stress. Medications utilized may include beta blockers (reduces the hearts workload), nitroglycerine (nitrates to increase oxygen through additional blood supply) and calcium channel blockers (increase blood flow). Invasive procedures employed to treat this condition would include Coronary Artery Bypass Graft (CABG) or Percutaneous Transluminal Coronary Angioplasty (PTCA) to increase blood flow to the heart muscle.

Angioplasty

This procedure is also called a "PTCA" or "Percutaneous Transluminal Coronary Angioplasty" which means that the procedure was done through the skin on a blood vessel that was an artery and that it was performed to reconfigure it. The PTCA is employed to open blockages in arterial vessels that have been compromised by diseases such as Coronary Artery Disease. This is a progressive disease in which plaque gathers within the arterial lumens to a degree to where there is less blood delivered to the heart. If left untreated the ischemia created by this disease could lead to myocardial damage through necrosis (cellular death). The PTCA is a minimally invasive procedure that utilizes a catheter tube with a balloon at the end to press built-up plaque back against the artery wall to open the vessels lumen for blood to flow through. The catheter tube is inserted into an artery, commonly the femoral artery in the groin area and guided with the aid if a viewing screen to the already established blockages in the coronary arteries. When the balloon is at the location of the blockage it is inflated to press the plaque back against the artery wall. After this is accomplished the air is released from the balloon and the catheter is removed or moved to another location to repeat the procedure. This procedure is also utilized to deploy stents which are drug treated wire mesh tubes that are delivered to collapsed or weak locations in vessel walls. These mesh tubes are placed over the catheter balloon and fed through the artery to the compromised location where the balloon is inflated and the mesh tube is opened against the artery wall. The catheter balloon is then deflated and is removed leaving the stent in the location where it was opened. This mesh tube provides structural integrity to ensure that the vessel remains open for blood flow. As with any invasive procedure when the body is entered there is the risk of complication and although the chances of complications arising as a direct result of this procedure are minimal, they are present. Some of these complications may include:

* dissection or tearing of the artery being re-opened.
* disturbance of an embolism that was present in an artery that could travel through the cardiovascular system and precipitate a heart attack.
* the possibility of inducing a fatal arrhythmia.
* allergic reaction to the dyes or drugs used during the procedure.
* bruising or infection at the site of the insertion.
* retroperitoneal bleeding into the groin area from the entry site of the catheter (if the entry site was the femoral artery located in the leg)

The Angioplasty is a minimally invasive procedure that is normally done following negative results or from the exploratory cardiac catherization procedure. The PTCA's purpose is to reconfigure narrowed areas within the coronary arteries after their locations have already been established. This is also referred to as revascularization therapy. Failed attempts at the PTCA procedure have the potential to progress into a more invasive surgical procedure such as Coronary Artery Bypass Graft (CABG). This procedure surgically reconstructs arterial passageways to bypass blocked sections of blood vessel to re-route the delivery of blood around the blockage.

Heart Rates

Sinus Rhythm	60 - 100 BPM
Sinus Bradycardia	Less Than 60 BPM
Sinus Tachycardia	100 - 160 BPM
Atrial Fibrillation:	
Controlled	<100 BPM
Uncontrolled	>100 BPM
Atrial Flutter:	
Controlled	< 100 BPM
Uncontrolled	> 100 BPM
Junctional	40 - 60 BPM
Junctional Bradycardia	20 - 40 BPM
Accelerated Junctional	60 - 100 BPM
Junctional Tachycardia	Greater Than 100 BPM
Idioventricular	20 - 40 BPM
no activity above the ventricles	Accelerated 40 - 100 BPM
Ventricular Tachycardia	150 - 250 BPM
Slow Ventricular Tachycardia	100 - 150 BPM
Torsade de Pointes	150 - 250 BPM
PAT / PSVT / SVT	160 - 250 BPM

The Hearts Intrinsic Rates

Sinoatrial Node	60 - 100
AV Junction	40 - 60
Ventricles	20 - 40

Arrhythmia

The arrhythmia, also referred to as "dysrhythmia," is an abnormal heart rhythm that has been induced by malfunctions or disturbances within the hearts conduction system. Under normal circumstances the hearts conduction system functions by a process called automaticity, which is a process by which the cells located within the heart have the ability to create electrical impulses independently and without the aide of outside stimulation. The initial electrical impulse originates in a group of specialized cells called the sinoatrial node that is located in the right atria or upper right chamber of the heart. This cell cluster has the potential to generate electrical impulses at a rate faster than the cells located further through the conduction system, making it the hearts natural pacemaker. As the electrical impulse travels through the conduction system it controls the depolarization of the other cells along the pathway, causing them to respond to the electrical stimulation as the impulse is transmitted from one part of the myocardium to another. In the normal cardiac cycle the electric impulse begins at the site of the sinoatrial node and travels along a prescribed conduction pathway until it ends at the purkinje fibers within the ventricles. As it is transmitted from the sinoatrial node to the purkinje fibers it induces prescribed rhythmic muscular responses in the form of contractions along its route, precipitating the movement of oxygen-rich blood through the heart and to the lungs and body. When disturbances arise such as an arrhythmic event the impulse transmission from one part of the myocardium to another is altered, compromising the hearts ability to function in synchronization. This action can adversely affect the entire body. The disorganized activity created by the abnormal cardiac behavior can diminish the volume of oxygen-rich blood that is delivered to the body and cause the organs cells to suffer damage or death through a degenerative process referred to as necrosis. Depending on the severity of the arrhythmia and the amount of damage inflicted by the ischemic atmosphere created, this process can happen in minutes or over a longer period of time. Arrhythmias can be induced by different factors such as a dysfunctional node, diseased myocardial tissue, certain cardiac medications, prior heart attack, or cardiac ischemia. Depending upon the factor that precipitated the event it can originate in different parts of the myocardium. A few examples of these areas would be:

SA Node This sinoatrial node is also referred to as the hearts natural pacemaker. This is a group of specialized cells/tissue that is the point of origin for the electrical impulse that begins the entire cardiac cycle. Malfunctions of this node can create problems that range from conduction delays which are referred "sinus pauses" or "sinus exit blocks" that can present for time durations that can last from a matter of seconds to complete failure of the generation of the electrical impulse that can result in death. This complete failure is referred to as asystole which is an event that stops all blood circulation to the body, causing death to its organs in minutes.

AV Junction The atrioventricular junction is comprised of the atrioventricular node and the bundle of his. It is at this point in the conduction cycle that the electrical impulse is transmitted from the atria to the ventricles.

Diseases or injury to these electrical impulse conduction sites compromise their ability to perform their function effectively, causing a conduction delays or complete blockage of the impulse. A few of the blocks associated with malfunctions of this junction are referred to as the atrioventricular blocks; also referred to as the AV blocks. These blocks are classified according to their characteristics. Examples of these would include:

* **1st degree AV block** This is considered more of a conduction delay due to the fact that all electrical impulses do make it to the ventricles. The AV block is measured from the beginning of the P-wave or the start of the atrial depolarization to the start of the QRS or ventricular depolarization, which is referred to as the PR interval; with the severity calculated according to its time duration as follows:
  * 0.12-0.20 normal
  * 0.21-0.24 considered slight
  * 0.25-0.29 considered moderate,
  * 0.30 and greater severe
* **2nd degree AV block "type 1"** This block is also referred to as "Mobitz 1" or "Wenckebach." This occurs when diseased tissue within the atrioventricular node causes a gradual lengthening of the PR interval until there is a dropped ventricular complex. It is seen commonly as a transient arrhythmia that appears for brief episodes that enters into and out of another base rhythm.
* **2nd degree AV block "type ll"** Second degree type two also presents with dropped QRS complexes and is commonly observed to consist of two P-waves per one QRS per cardiac cycle. This is referred to as 2:1 conduction. The distinguishable difference between 2nd degree type 1 and 2nd degree type 2 is that in 2nd degree type 2 there is no gradual lengthening of the PR interval to forewarn of the impending dropped ventricular beat. This arrhythmia can be transient in nature or sustained.
* **3rd degree AV block** Also called "complete heart block," 3rd degree AV block is an arrhythmia where there is a complete lack of communication between the atria and the ventricles at the site of the AV Junction. This arrhythmia can present irregular, will have a variable PRI and will possess a ventricular heart rate between 20-40 beats per-minute due to the intrinsic conduction rate of to the purkinje fibers which are are providing the electrical impulse to depolarize the ventricles (automaticity).

Another form of an arrhythmia is atrial fibrillation which is also called "A-Fib." This is the result of disorganized electrical activity within the atria created by rogue impulses firing erratically outside of the hearts normal conduction pathway. This disorganization causes the atria to fibrillate or quiver instead of its normally controlled contractions. Atrial fibrillation is an irregular rhythm with no discernible P-wave that is considered controlled with a heart rate below 100 beats per minute and uncontrolled with a heart rate greater than 100 beats per minute. This arrhythmia can be sustained and lasting for a long period of time, paroxysmal in nature entering and leaving another rhythm in short episodes, or chronic and occurring frequently. One of the greatest concerns related to atrial fibrillation is the formation of clots within the hearts chambers due to the lack of blood movement. These clots can enter and travel through the circulatory system with the potential of inducing a heart attack or stroke.

Did You Know?

There are one and one-half million heart attacks annually in the U.S., with five-hundred thousand resulting in death.

Ashman Phenomenon

Ashman Phenomenon is a wide complex ventricular conduction that is similar in appearance to a premature ventricular contraction or "PVC." This phenomenon is induced by a form of triggered automaticity that is the result of the stimulation of an electrical impulse from a premature cardiac cycle that has entered into the refractory period of a previous cardiac cycle that was prolonged (a prolonged beat followed by a premature beat). This action triggers an abnormal ventricular response when it attempts to depolarize vulnerable cells present in an incomplete ventricular repolarization phase. The beat will be aberrant due to its occurrence within the cellular refractory period and will possess a wide complex and a right bundle branch block due to the longer refractory period of the right ventricle over the left ventricle. Because of it's disposition to cause cardiac cycle irregularity, Ashman Phenomenon is present mainly in atrial arrhythmias such as atrial fibrillation, supraventricular tachycardias and rhythms that produce frequent atrial ectopy. The primary tool used by Physicians to evaluate the heart rhythm to form the diagnosis of this anomaly is the electrocardiogram or "EKG." This test records the electrical activity of the hearts conduction system as the electrical impulse travels its route from the sinoatrial node to the purkinje fibers. This phenomenon was first discovered in the year 1947 by J.L. Gouaux and R. Ashman through the observation that ventricular aberrant beats are more pronounced when they present in the atrial fibrillation rhythm while occurring in the refractory period. Multiple beats of this phenomenon occurring in a run may be confused with ventricular tachycardia which presents to be similar in appearance.

* Absolute Refractory Period – this phase of the refractory begins at the onset of the ventricular depolarization (Q-wave) and extends half way into the T-wave or mid way into the repolarization period of the cardiac cycle. This phase of the cycle presents cardiac cells in a state of flux at which they have not sufficiently recharged. These cells will not respond to premature electrical stimulation regardless of its intensity.

* Relative Refractory Period – This period corresponds with the cardiac cycle from mid way into the repolarization or at the point half way through the T-wave, down the T-wave slope to a point before the T-wave merges with the isoelectric baseline (completion of the repolarization phase). This phase of the refractory period possesses cardiac cells that have sufficiently recharged to a level that makes it possible for an electrical impulse with enough intensity to induce triggered automaticity and initiate a fatal rhythm.

* Supernormal Refractory Period – The supernormal refractory period begins as the cardiac cells within the ventricles reach their completion. This is located on the EKG strip where the T-wave returns to the baseline. In this state of recharge the cells will respond to premature electrical stimulation of normal intensity.

Did You Know?

The first case report of atrial fibrillation was recorded in the year 1907 by Aurthur Cushny who was a professor of pharmacology at the University College of London. The patient was a female who underwent surgery on an ovarian fibroid tumor and developedan irregular heart beat with a heart rate that reached 160 beats per minute. This event occurred three days post-op.

Asystole

Asystole is described as the total stoppage of all electrical activity by the sinoatrial node to stimulate myocardial contractions. This creates a situation where there is an absence of oxygen rich blood supply to the body, causing tissue and organs to become permanently damaged or die depending on the duration of time without the presence of oxygen. This diagnosis must be confirmed in two separate leads on a heart monitor to rule out faulty leads or fine ventricular fibrillation. Asystole can be classified into two categories, primary asystole and secondary asystole. The primary form of asystole occurs when the hearts natural pacemaker (sinoatrial node) can no longer metabolically produce an electrical impulse. This leaves the heart in a depolarized state that eliminates the option of utilizing defibrillation to reverse or correct the problem. A couple of possible conditions that could precipitate primary asystole could include diseases such as coronary artery disease which over time affects the heart through ischemia, or a natural deterioration of the hearts conduction system that occurs with age. Secondary asystole happens as the result of complications related to other serious medical issues that can include stroke, heart attack (MI), suffocation, hyperkalemia, respiratory failure, pulmonary embolus, or a deteriorated state of the ventricular fibrillation or the ventricular tachycardia arrhythmias.

Non-medicinal treatment for this condition would include steps to protect the body's tissue and organs such as intubation to deliver oxygen, the establishment of vascular access to deliver medications, the initiation of CPR (cardio pulmonary resuscitation) to provide minimal circulation to the body's tissue and organs and the implementation of transcutaneous pacing to deliver electrical impulses to manage the heart. Medicinally, drugs such as epinephrin, atropine and vasopressin are administered to stimulate muscular activity. The utilization of defibrillation would not produce a positive response due to the fact that the heart is in a depolarized state, but may also worsen the situation by reducing the possibility of regaining the hearts rhythm. This event has a poor prognosis with a questionable probability of a chance of resulting neurological damage.

Did You Know?

In 1931 the use of exercise was utilized to intentionally induce angina pectoris by Francis Wood and Charles Wolferth. The electrocardiograph (EKG) findings created during the exercise sessions were compared to normal tracings for comparison. These tests were discontinued due to the dangers of inducing indiscriminate angina.

The first Holter Monitor was invented in the year 1949 by Dr. Norman Holter. It consisted of a pack that was worn on the individuals back that weighed 75 pounds that could record and transmit the heart rhythm of its wearer. Today, the Holter Monitor is greatly reduced in size to that of a vest that records cardiac activity while the patient ambulates during their normal daily routine.

Atrial Fibrillation

This arrhythmia is the culmination of disorganized electrical impulses of different amplitude that precipitate the atrial myocardium to respond at a rate of 300 beats per minute or greater. The defined atrial depolarization normally seen in sinus rhythm is replaced with erratic atrial quivering that presents with no discernible P-wave on the EKG tracing. Multiple electrical impulses that pass through the atrioventricular node in an unregulated fashion activate the ventricles without organization, making the ventricular rhythm irregular. This abnormal electrical activity can also cause them to depolarize at a rate that can exceed 150 beats per minute, with the rate dependant on the stability of the atrioventricular node. The isoelectric baseline can be coarse as seen in a case of "new onset" atrial fibrillation, or fine as observed in a longer standing more established case of this arrhythmia. Atrial Fibrillation is an arrhythmia that is more frequently encountered among patients of advanced age, and it is estimated that:

* an individuals probability of becoming afflicted with this arrhythmia doubles every ten years after the age of fifty-five.
* this heart arrhythmia affects three to five percent of individuals that are sixty years of age.
* atrial fibrillation is prevalent in nine percent of those at or greater than the age of seventy-five.
* it is also estimated that the mortality rate from strokes in patients afflicted with this arrhythmia is two times greater than for those that are non-afflicted.

To manage this arrhythmia, coagulation and ventricular rate control need to be addressed. Due to the hearts inability to successfully complete a depolarization cycle and move a complete fill of blood from its chambers, the blood pools within the heart and creates a situation where coagulation (clotting) can occur. To accomplish successful anticoagulation a couple of the more frequently used forms of anti-clotting therapy may include:

* heparin – this substance is found within the different bodily tissue, mainly in the liver and acts as a anti-clotting agent by indirectly inhibiting thrombin which is in the blood and is critical for clotting. This is administered for a seven day period while the patients lab values are closely monitored.
* coumadin - this chemical prevents clotting by inhibiting the blood clotting factors that depend on vitamin K. This agent is also used to prevent future occurrences of blood clotting where there is a history of a prior heart attack, or the recurrence of clotting such as those related to a condition called "deep vein thrombosis".

Although successful blood anticoagulation can be accomplished using these agents, there is a fine line between the point that an individual is sufficiently anticoagulated and the risk of uncontrolled bleeding is breached. Therefore, continuous monitoring of the patients clotting factors through their lab work is necessary.

Among other problems encountered with this arrhythmia is the rapid heart rate associated with uncontrolled ventriclular response created by the erratic fashion in which electrical impulses are allowed to pass through the atrioventricular node. This rapid rate compromises the hearts ability to provide enough fill time for its chambers to receive a sufficient amount of blood to distribute to the rest of the body and to the lungs.

Unable to meet the body's demands for oxygen-rich blood due to its compromised blood supply, the heart works harder in an effort to fulfill the body's needs. This unmet myocardial oxygen demand eventually leads to ischemia and possible heart failure. To regain control of this rapid rate there are a couple of options. One of these options is the utilization of specific drugs that are administered for ventricular rate control, they are:

* Verapamil	**type**: calcium channel blocker	class: acute 1V
* Cardizem	**type**: calcium channel blocker	class: acute 1V
* Digoxin	**type**: beta blocker	class: acute 1V
* Propranolol	**type**: beta blocker	class: acute IV

Other options available for controlling rapid heart rates may include:

* the introduction of a Implantable Cardioverter Defibrillator or "ICD." This device is surgically implanted into the patients chest to act as an on-board cardioversion unit to regain control in situations involving sudden episodes of rapid atrial fibrillation.
* the implantation of a ventricular pacemaker to control and manage the ventricles
* a radiofrequency ablation of the atrioventricular node

It is important to research the history of the atrial fibrillation arrhythmia in the patient. Knowing how long it has been present in the patient and how much damage to the heart (ex. enlarged left atrium) has resulted is information needed to decide whether a conversion back to a normal sinus rhythm should be attempted and is safe and if it shows that there is a good possibility of maintaining the rhythm change once it is accomplished. This conversion of the arrhythmia back to normal sinus rhythm is referred to as a "cardioversion" and can be accomplished in either of two ways. One method is the utilization of medications, some of which may include:

* Amiodarone	**type**: antidysrhythmic	class: 111
* Rythmol	**type**: antidysrhythmic	class: 1C
* Norpace	**type**: antidysrhyrhmic	class: 1A
* Tambocor	**type**: antidysrhythmic	class: 1C

The other option available to be utilized in converting atrial fibrillation back to normal sinus rhythm is a procedure that is referred to as "electrical cardioversion." Electrical cardioversion is a procedure that manipulates the conduction system by introducing a direct current electrical shock that depolarizes the myocardium, creating an interruption in the erratic electrical impulses associated with the atrial fibrillation arrhythmia. The temporary window induced by this procedure provides an opportunity for the sinoatrial node to reset and re-establish itself as the hearts dominant pacemaker. This procedure is more successful in converting and maintaining sinus rhythm in cases that are less than forty-eight hours old that are referred to as "new-onset" atrial fibrillation. This procedure should not be performed on patients that have been in this arrhythmia for more than 48 hours until after they have been checked for the presence of atrial clotting. Cardioversion, whether electrical or medicinal in modality performed in the presence of atrial thrombi may precipitate the development of emboli that can travel through the bloodstream causing further medical complications or death.

Terminology used to describe the various forms of atrial fibrillation can include:

* controlled – possessing a ventricular heart rate under 100 beats per minute
* uncontrolled – possessing a ventricular heart rate greater than 100 beats per minute (a ventricular heart rate greater than 150 beats per minute is normally referred to as rapid)
* coarse – refers to the amount of rough and erratic electrical behavior present in the isoelectric baseline. A coarser isoelectric baseline may indicate a newly developed onset of the arrhythmia.
* fine – this description would be the opposite of that explained in coarse atrial fibrillation. An isoelectric baseline with minimal erratic electrical activity present is usually identified as being associated with a more established, longer standing case of this arrhythmia.
* PSVT – this arrhythmia in a transient, uncontrolled/rapid form, may be described as a paroxysmal supra-ventricular tachycardia. This would be defined as abnormal electrical activity above the ventricles that precipitates rapid ventricular response that can be seen to start and stop in short episodes. PSVT is considered a generic term to describe multiple atrial induced arrhythmias that possess the properties that give them the capability to increase the ventricular heart rate between 160 - 250 beats per minute.

Note: this arrhythmia possesses an irregular rhythm that can be confused with the arrhythmias "wandering atrial pacemaker" (WAP), and "multi-focal atrial tachycardia" (MAT). These rhythms are also atrial induced and display an irregularity that closely resembles that of atrial fibrillation. It is necessary to confirm the atrial fibrillation rhythm by examining the EKG tracing for P-waves that possess various morphologies consistent with WAP and MAT.

Did You Know?

Of the heart attack victims that do not receive prompt medical intervention from medical staff within a hospital, fifty percent will die within the first hour.

There is approximately sixty-thousand miles of blood vessels that run through your body with enough blood pressure within your arteries to send a stream of blood five feet into the air if they are opened.

Atrial Flutter

Atrial Flutter is a reentrant atrial arrhythmia that is precipitated by the looping motion of a premature electrical impulse that had originated from within the right or left atria. This arrhythmia sustains its motion through triggered automaticity precipitated by the electrical impulse that had entered into vulnerable atrial myocardial cells that have not yet completely repolarized. This action triggers a recurring atrial myocardial response in the form of a fluttering motion. Visually, this arrhythmia possesses very distinguishable waves that are referred to as F-waves that have a saw-tooth appearance and are generally larger and more squared than the normal P-waves. The atrial rate can reach between 250-400 impulses per minute, with the ventricular rate dependant on the extent of the block. Atrial Flutter can be regular when seen in conduction cycles such as 2:1 which consistently produces two F-waves for each QRS complex, or it can be variable and displaying an irregular pattern where there is no uniformity to the number of F-waves per QRS complexes. This arrhythmia is classified as controlled with a heart rate less than 100 beats per minute, uncontrolled with a heart rate greater than 100 beats per minute and is categorized as a form of supra-ventricular tachycardia (SVT) when heart rate reaches and sustains 160 beats per minute or greater. Atrial Flutter can precipitate a variety of medical problems, such as:

* the formation of clots within the hearts chambers resulting from lack of blood movement due to poor circulation. These clots can then travel through the circulatory system and become lodged within an artery, inducing a heart attack or stroke. To lessen the occurrence of this event anticoagulants such as heparin and coumadin are used to prevent coagulation.
* syncope and pre-syncope episodes can become a factor when this arrhythmia precipitates an extremely low heart rate that is referred to as bradycardia. Oxygen deprivation to the brain causes the individual to experience lightheadedness or fainting.
* angina can occur due to the diminished volume of oxygen-rich blood delivered to the heart muscle. This is chest pain that can radiate to the right arm, shoulder and jaw, or present as a pressure/tightness in the chest area. Angina is commonly associated with and is induced by oxygen deprivation.
* the blood in the veins returning to the heart becomes backed up by the lack of force from the weakened perfusion causing the fluid portion of the blood to accumulate in the surrounding tissue and lungs. This creates a condition that is called edema.

Some of the medical conditions or diseases that may contribute to the onset of this arrhythmia can include heart failure, pulmonary embolism, hypoxia, chronic obstructive pulmonary disease, hyperthyroidism, or a diseased mitral valve.

There are a couple of options available as interventional treatment, they would include:

* medications such as verapamil to lengthen the atrioventricular conduction time and anti-coagulation therapy to prevent clotting.
* cardioversion to intentionally depolarize the heart through a timed electrical shock. This causes an interruption in the conduction system that gives the sinoatrial node the opportunity to re-establish itself as the hearts dominant pacemaker.
* radiofrequency catheter ablation is a procedure that is employed to destroy tissue that may be inducing the arrhythmia.

Atrial Septal Defect (ASD)

Atrial Septal Defect, also referred to as "ASD," is a congenital defect that creates a hole in the septum between the right and left atria. This condition can lead to a multitude of cardiac related medical problems which may include:

* allowing blood to flow from atria to atria, normally from the left to the right due to the increased pressure present in the left atria. This forcing of blood from the left to the right allows blood that has already been re-oxygenated to cycle through the lungs to be processed over again, diminishing the blood supply to the body.
* can create a condition that is referred to as dilated cardiomyopathy which is the stretching of the hearts chambers due to volume overload, eventually debilitating its ability to function through loss of flexibility and contractability.
* the blood volume overload created in the right atrial chamber causes a back-up putting excessive pressure on the pulmonary arteries precipitating a condition called pulmonary hypertension, resulting in a fluid build-up within the lungs or pulmonary edema.
* when the pressure gets excessive the blood can then reverse shunt sending waste blood returning from the body into the right atria to be mixed with the blood in the left atria en-route to the left ventricle to be sent to the body without being re-oxygenated. This condition can inflict serious damage to the body's organs and tissue through a condition called hypoxia/ischemia, which is the damage or death of the body's cells caused by necrosis.
* The additional workload on the heart to provide a sufficient amount of blood that has not been re-oxygenated to satisfy the body's needs can lead to conditions such as hypertrophy which is a thickening of the left ventricle chamber walls causing loss of contractability and diminished perfusion.
* can result in irreversible Congestive Heart Failure

Symptoms related to this condition my include hypertension, shortness of breath, labored breathing, fatigue, and dyspnea, creating the inability to perform everyday tasks. This condition is corrected through cardiac surgery with the implementation of a patch or graft to block the passage from atria to atria.

Did You Know

It is estimated by the FDA that in the year 2006, 25% of the reported heart attacks (roughly around one-hundred and seventy-five thousand) were "silent" and would display no symptoms. Unfortunately, heart attacks evolve and cause greater myocardial injury over time which gives these unrealized cases more time to inflict damage before medical intervention is obtained.

Automated Implantable Cardioverter Defibrillator

The Automated Implantable Cardioverter Defibrillator, which is also referred to as an "AICD" or "Defibrillator," is a device that is designed to deliver an electric shock to correct or terminate arrhythmias to restore normal sinus rhythm. Research began in the year 1969 with a team at Sinai Hospital in Baltimore that was conducted over an 11 year period to design a means to correct heart arrhythmias when and where they occur; without the aid of antiarrhythmic drugs. The first AICD patient implant was performed in the year 1980 at John Hopkins Hospital by Dr. Levi Watkins. The AICD is small in size measuring a little over 12mm thick with a weight of approximately 70 grams and is designed to perform different functions such as defibrillation, overdrive pacing and pacing, This device is surgically implanted subcutaneously in the upper left chest in a pocket located just below the skin and transmits its electrical impulse to the myocardium through a wire or "lead" that is fed through a vein that leads to the heart where it is positioned in the ventricle. There are a couple of different methods utilized by the computer in the AICD when interrogating arrhythmias. One method measures heart rates between the atria and the ventricles to determine the arrhythmias origin. This function is called "rate discrimination." Another function is the evaluation of rhythm regularity. With most atrial arrhythmias being irregular this method can differentiate between atria and ventricular origin. After the type of arrhythmia has been determined the proper function can be employed to treat it. Some of the various functions would include:

* Defibrillation – the defibrillation mode is used for arrhythmias such as Ventricular Fibrillation, Ventricular Tachycardia and Torsade de Pointes. This function is programmed to induce an intentional pause in the conduction system (depolarize) to allow the sinoatrial node to reset. This objective of this action is to terminate the arrhythmia and present a window for the heart to restart in normal sinus rhythm.
* Overdrive Pacing is utilized in correcting arrhythmias such as Ventricular Tachycardia before they can deteriorate to Ventricular Fibrillation. This function delivers impulses at a rate greater than the rate of the arrhythmia to regain control of the conduction system. The rhythm is then controlled by the AICD which gradually reduces it to a controlled rate.
* Pacing mode is programmed to maintain a preset heart rate in situations where the heart rate chronically drops to a level that induces a symptomatic response. The unit is pre-programmed to a preset heart rate that it will not allow the heart to drop below.

Another function that can be incorporated into the design of the AICD is bi-ventricular pacing. This feature is utilized to manage conduction systems that have been compromised by age or diseases such as congestive heart failure. bi-ventricular pacing which is also referred to as "resynchronization therapy," is programmed to deliver electrical impulses through two separate lead wires to both the right and left ventricles simultaneously to re-establish synchronization to the ventricular contractions.

This same technology that is used in the automated implantable cardioverter defibrillator (AICD), is available in the automated external defibrillator or "AED." This device is portable and is designed to be carried to the scene of an emergency by EMT's and other emergency personnel. This device is designed to sense and read heart rhythms, then to advise the user of the proper course of action to be taken.

Atrio-Ventricular Dissociation
(AV Dissociation)

AV Dissociation is a disturbance within the hearts conduction system that causes there to be a separation between the ventricular and atrial function that precipitates unsynchronized myocardial contractions. This abnormal behavior can shorten chamber refilling time causing diminished blood flow to the body and its organs that in severe cases can induce symptomatic responses that may include palpitations, dizziness, fatigue and shortness of breath that can occur both at rest and during exercise or other forms of physical activity. AV dissociation can present in different forms that can possess varying levels of severity. A few of these variances could include Isorhythmic AV dissociation which occurs when the impulse discharge rate from the site of the sinoatrial (SA) node (hearts natural pacemaker) which is located in the upper right atria becomes bradycardic and drops to a rate that is lower than that of the intrinsic rate of the AV junction which is 40-60 beats per minute. When the decrease in electrical impulse discharged from the SA node occurs, the escape mechanism within the AV junction functions independently and produces its own impulses through cellular automaticity. These impulses are then conducted to the ventricles at a rate between 40-60 beats per minute. This creates a situation where the atrial chambers and the ventricular chambers can function at approximately the same rate of contractions per minute, but due to the fact that the compromised SA node impulse rate is slower than the intrinsic rate of the AV junction the ventricular chambers are contracting minimally faster than the atrial chambers. The isorhythmic form of AV dissociation produces two separate rhythms contracting independently at approximately the same rate with regularity. Another form of AV dissociation is referred to as Interference AV Dissociation and can be related to arrhythmias such as 2nd degree atrioventricular (AV) block. Due to its characteristic behavior, this form of dissociation intermittently blocks the progression of the conduction of the electrical impulse from the atria to the ventricles, causing there to be more atrial depolarizations than ventricular depolarizations creating a faster atrial rate. "Complete Heart Block," also called "3rd Degree AV Block," is described as the complete failure at the site of the atrioventricular (AV) node to conduct electrical impulses from the atria to the ventricles. This prompts the automaticity within the ventricular purkinje fiber system to become the escape mechanism to prevent ventricular standstill. The complete blockage of communication from this form of AV Dissociation will cause there to be two separate and distinct rhythms. The atrial contractions will be regular with a rate normally between 60-100 beats per minute and the ventricular contractions will be irregular with a discharge rate between 20-40 beats per minute, each in accordance with their own intrinsic electrical impulse discharge rate. The AV Dissociation arrhythmia is diagnosed with the utilization of a test that is referred to as the electrocardiogram (EKG).

<u>Intrinsic electrical impulse heart rates</u>:

Sinoatrial (SA) node	60-100 beats per minute
Atrioventricular (AV) Junction	40-60 beats per minute
Purkinje fiber system	20-40 beats per minute

Bi-Ventricular Pacemaker

The bi-ventricular pacemaker design came about in the year 2001 (with the addition of an ICD in 2002). It's purpose is to reestablish synchronized contractions of the hearts ventricular chambers, giving its implementation the title "Cardiac Resynchronization Therapy" or "CRT." These units are used in approximately thirty-five percent of the five-million patients yearly that are stricken with irreversible heart conditions such as congestive heart failure. In the normal function of the heart, the right ventricle and left ventricle contract in synchronization to provide a firmness and pressure to the walls of the left ventricle during depolarization. This produces a strong catalyst during perfusion when the left ventricle contracts to move blood throughout the body. In many chronic heart failure patients the ventricles lose their ability to function together due to abnormalities in the hearts conduction system, resulting in unsynchronized contractions of the ventricular chambers. These unregulated depolarizations causes the septum between the right and left ventricle to flex when the left ventricle contracts, impairing its ability to produce enough force to adequately pump blood to the body. The diminished cardiac output created by the weakened perfusion makes the heart work harder to make allowances for the lower volume of oxygen-rich blood being delivered to the body's organs and tissue, eventually causing the heart muscle to become weakened and lose its flexibility. The bi-ventricular pacemaker unit employs three leads, one each for the right and left ventricle and a third that is placed within the atrium. The computer is programmed to deliver an electrical impulse to stimulate both the right and left ventricle simultaneously, causing them to contract together. This benefits the individual by not only increasing cardiac output, but by decreasing the additional workload on the heart. This device may be utilized as a single function "stand alone" unit, or can be integrated with an Implantable Cardioverter Defibrillator or "ICD" to perform dual functions. Statistics have shown that when configured in this dual role fashion, the death rate from complications brought on by these types of potentially fatal occurrences can drop by as much as forty percent.

Did You Know?

After the implementation of a pacemaker, certain information is made available through a process called "transtelephonic monitoring." By utilizing a special device, information that is gathered and stored within the pacemaker can be transmitted from the individual's residence to the physicians office. During this transtelephonic communication a pacemaker check occurs making it possible to evaluate the history and programming.

In the year 1947 the successful defibrillation of a 14 year old boy was performed during cardiac surgery by a cardiologist named Claude Beck. This success came after 6 previous attempts on other patients failed to generate a positive response.

Bundle Branch Block

The bundle branch block, which is also referred to as "BBB," is a blockage in the hearts conduction system that occurs in either or both of the bundle branches located below the atrioventricular node following the bundle of His where there is a separation to the right and left ventricles. These branches are referred to as the right bundle branch and the left bundle branch. Abnormalities or injuries present within these branches can cause the electrical impulse to negotiate a different route to arrive at the point in the purkinje fibers where it stimulates a muscular response (contraction). This precipitates the loss of synchronization between the contractions of the ventricles and causes the impulse to exceed the time duration that normally occurs within a healthy conduction system which is 0.06 to 0.10 hundredths of a second. On an electrocardiogram (EKG) tracing, there is the presence of a bundle branch block if there is a time duration greater than 0.12 hundredths of a second from the beginning of the QRS complex to the J-point where the tracing changes direction and returns to the isoelectric baseline after the S-wave. It is at this point that the depolarization ends and the heart muscle enters into the ventricular repolarization phase of the cardiac cycle (T-wave). This abnormality can present for no known reason (idiopathic) or may be congenital in nature, but can also be caused by events such as a myocardial infarction (heart attack), the presence heart disease, injury that occurred during cardiac surgery, deterioration of the conduction system from the natural aging process, chronic obstructive pulmonary disease (COPD) or hypertension. The extent and seriousness of this condition is more pronounced in the left ventricular branch as opposed to the right ventricular branch and can range from a slight to severe. The bundle branch block can produce symptoms such as pre-syncope (feeling of dizziness or an impending fainting episode), syncope (fainting), a bradycardic heart rate, or the onset of heart arrhythmias that in extreme cases may result in sudden cardiac death. Diagnosis of this condition is routinely done with the use of a test called an electrocardiogram or EKG which records the hearts electrical impulse as it travels through the heart from its inception in the sinoatrial node (SA node) to its completion in the purkinje fibers located within the ventricles. This test produces a tracing that is used to measure the time durations at different intervals throughout the conduction system to ensure that the measurements do not exceed the set guidelines known to be present in a normal sinus rhythm. Measurements greater than this set limit indicate that there is a prolonged time duration experienced at that point in the conduction cycle and establishes the presence of a block. Once a bundle branch block has been established, there are a variety of tests to help a physician to determine the actual cause and to decide on the proper treatment. Treatment in most cases is not necessary due to the fact that the bundle branch block generally produces not symptoms, however, depending upon the severity of the situation can range from lifestyle changes to the introduction of medication to treat a condition that may be precipitating the block. Examples of medicinal treatment could include the implementation of blood pressure control drug therapy and medications to reduce symptoms caused by diseases such as congestive heart failure. In extreme cases a pacemaker can be surgically implanted into the individual's chest to manage the hearts conduction cycles.

Electrolytes

An electrolyte is an electrical charged ion that conducts and transmits electrical current to positive and negative electrodes. Cardiac electrolytes maintain voltage across the membrane of cardiac cells and facilitate the transmission of electrical current to other cells. Electrolytes such as calcium to provide cardiac cellular automaticity, magnesium to facilitate the transmission of the electrolytes across the membranes of the cells and potassium to maintain cellular neutrality enable the cells to function independently without outside stimulation. These impulses produce muscular responses in the form of contractions which cause the hearts chambers to squeeze, providing a supply of oxygen-rich blood throughout the circulatory system to the body and its organs. The amount of these electrolytes that are present in the blood is controlled by the amount taken into the body in food and drink in relation to what is lost from the system through urine, sweat and feces.

Sodium Na^+	135 - 145 mEq/L
Potassium K^+	3.5 - 5 mEq/L
Calcium Ca^{2+}	8.5 – 10.5 mg/dl
Chloride Cl^-	96 - 106 mEq/L
Magnesium Mg^{2+}	1.8 – 2.5 mEq/L
Phosphate PO_4^{2-}	2.5 – 4.5 mg/dl

A few EKG characteristics caused by an imbalance of electrolyte levels:

Hypokalemia (low potassium level)
- elevated U-wave
- flattened T-wave
- ST segment depression

Hyperkalemia (elevated potassium level)
- elevated / tented T-wave
- widened QRS complex
- ST segment depression
- flattened P-wave

Hypocalcemia (low calcium level)
- prolonged QT interval
- lengthened ST segment
- heart arrhythmias

Hypercalcemia (elevated calcium levels)
- shortened QT interval
- heart blocks

Hypermagnesemia (elevated magnesium level)
- widened QRS complex
- prolonged QT interval
- bradycardic heart rate
- heart blocks

Cardiac Arrest

Cardiac Arrest, also called "Circulatory Arrest" or "Cardiopulmonary Arrest," is an event that occurs when the heart muscle is compromised to a degree that it stops pumping blood to the body's organs and presents no arterial blood pressure. Without circulation of oxygen-rich blood to feed the body, cells and tissue can be irreversibly damaged and begin to die within three to four minutes of the onset through a process called necrosis. Death that occurs as the result of this event is referred to as "sudden cardiac death" and is considered a natural death that was the result of the natural progression of diseases such as coronary artery disease. It is estimated that as many as three-hundred thousand individuals die due to cardiac arrest per year due to the fact that up to eighty percent of those who suffer this fatal cardiac event are unable to receive immediate medical intervention. Initial treatment given to the individual outside of a hospital setting would include basic life support (BLS) with cardiopulmonary resuscitation (CPR) started to maintain blood and oxygen circulating to the body's organs until proper medical intervention can be employed. Intervention within a hospital setting would follow a protocol to institute an advanced cardiac life support (ACLS). Once the determination is made as to the reason for the arrest, the proper medical treatment in the form of defibrillation or medication can be given. The origin that precipitated the arrest must be identified and treated accordingly. Arrhythmias such as Ventricular Fibrillation and Ventricular Tachycardia are "shockable" rhythms that respond to and can be reversed by electrical stimulation. In these arrhythmias the introduction of electrical stimuli creates an interruption in the conduction system which allows the sinoatrial node to regain control of the heart when it starts again. Situations that involve Asystole and Pulseless Electrical Activity (PEA) are not corrected by defibrillation. Asystole presents the heart in a depolarized state that is absent of and unable to generate electrical activity; the introduction a shock into a depolarized heart that cannot generate its own electrical activity would serve no purpose. Pulseless Electrical Activity or "PEA" is a problem that presents the myocardium active with electrical activity but absent of the mechanical or muscular activity. This condition displays the presence of electrical activity on a cardiac monitor but the absence of a carotid pulse. In these two instances drugs such as atropine, potassium, amiodarone, magnesium, epinephrine (adrenalin), or calcium are administered to chemically enhance the hearts muscular activity. There are other factors that can induce a "secondary cardiac arrest", or a cardiac arrest that was induced as the result of event. A few of the situations that could contribute to the onset of this event would include:

* heart arrhythmias that cause extreme bradycardia. These arrhythmias create an ischemic atmosphere that starve the heart of oxygen rich blood, eventually causing failure.
* cardiac tamponade is a condition that puts an external constrictive pressure on the heart muscle that prevents oxygen-rich blood flow to its surface arteries, causing the heart to suffer through ischemia. Tamponade also restricts the expansion of the hearts chambers, compromising their ability to fully expand an receive a fill of new blood. This can sometimes be the result of trauma or pericarditis, which causes swelling or inflammation of the pericardial sac that surrounds the heart.

* sepsis which is a poisoned state in blood that has absorbed pathogenic microorganisms.
* hypoxia - hypoxia is the inadequate supply of oxygen to the heart muscle and other organs. This can be precipitated by problems with the carbon dioxide to oxygen exchange taking place with the lungs, or it can be related to ischemia. Ischemia occurs when there is a diminished supply of oxygen-rich blood to the heart due to diseases such as Coronary Artery Disease or CAD, where built-up of fatty tissue and plaque within the arteries have narrowed the passageways.
* electrocution – the introduction of electrical interference from an outside source that enters the body and depolarizes the myocardium.
* hypovolemia - which is a low blood circulating volume due to blood loss. In this situation transfusions and IV fluids are introduced into the body to increase the circulating volume.
* hypokalamia - a low potassium level makes the individual susceptible to arrhythmias such as Polymorphic Ventricular Tachycardia (Torsade de Pointes). This presents on a cardiac monitor with flat T-waves and pronounced U-waves, and is considered a contributor to a conduction anomaly called Long QT Syndrome.
* hyperkalemia - a high potassium level that can be seen in patients with end-stage renal disease. This presents on a cardiac monitor with a high and peaked T-waves.
* hypoglycemia which is a low glucose level in the blood or hyperglycemia which is a elevated glucose level in the blood
* toxicity - poisoning from chemicals, or the overdose or lethal combination of either prescribed, over the counter, or illegal drugs.
* thrombosis which precipitates a myocardial infarction through the blockage of a blood vessel in the heart or lung caused by a thromboembolism. This can be induced by a blood clot, piece of tissue, or air bubble that has traveled through the circulatory system and become lodged within an artery. This is generally treated with thrombolytics or TPN, which is also referred to as "clot busters," to dissolve the blockage and reopen the vessel.
* hypothermia occurs when the body's core temperature drops to and maintains a temperature of 95F or lower. Bodily core temperatures below 86F excludes the use of defibrillation, making cardiopulmonary resuscitation necessary until bodily temperature allows defibrillation to be utilized.
* a tension pneumothorax occurs when air build-up within the pleural cavities causes a positional change that creates a reduction or stoppage in the flow of blood in a great vessel that supplies the heart such as the superior vena cava, limiting or stopping the amount of blood flow back to the heart.

Did You Know?

A thin sac that is called the pericardium surrounds the heart. This protective envelope is made of a tough tissue that acts as a buffer to protect the heart from damage caused by contact between itself, the lungs and the chest wall. There is a fluid that is discharged between its smooth interior wall lining and the heart to act as a lubricant to protect it from friction caused by rubbing.

Cardiac Catherization

The cardiac catherization is a minimally invasive procedure that is performed to gather information about the heart and the coronary arteries. This procedure allows the physician to examine the heart as it is beating and functioning and is also useful in forming a diagnosis for the presence of medical conditions such as coronary artery disease. During this procedure various tasks are performed such as the taking of x-rays, the measurement of the blood pressure within the heart and its major arteries and the taking of blood for test samples. The Physician will inject a contrast dye through the catheter tube into the coronary arteries to carry out a test called the coronary angiogram. This test allows the Physician to observe blood flow through the arteries and heart, evaluate the function of the valves and assess the movement of the hearts chambers during the contraction (depolarization) phase of the cardiac cycle. The catheter is then repositioned and the dye is used to perform a left ventriculogram, which is a test that is done to evaluate the volume of blood that is sent to the aorta by the left ventricle. Following this procedure the Physician will review the results to see if the tests may reveal abnormalities such as regurgitation through the hearts valves, insufficient blood volume delivered to the aorta, or pulmonary hypertension.

To carry-out the cardiac catherization, the Physician makes a small cut in an artery to be used as an entry point for the catheter. The most commonly used blood vessel is the femoral artery located in the groin area, however, in certain situations the arm is selected as an alternate entry site. After the incision is made, a guide wire is fed into the artery to the heart and a thin tube called a catheter is introduced along the guide wire while the physician observes its progression on a fluoroscope monitor screen to watch for abnormalities or blockages. Although this procedure is considered safe with over one-million performed each year, there is the possibility as with any other medical procedure that complications may occur. Some of the medical problems that may be encountered can include:

* excessive bleeding from the incision made at the entry site or within the body such as a retroperitoneal bleed (bleeding backwards into the abdominal cavity).
* dissection of an blood vessel, which would cause this procedure to progress into a more invasive emergency cardiothoracic surgical procedure called the Coronary Artery Bypass Graft (CABG).
* complications from an infection at the site of the incison.
* the onset of an potentially fatal arrhythmia.
* the possibility of a blood clot, fatty tissue, or plaque breaking free by the movement of the guide wire or catheter precipitating a heart attack or stroke.

Did You Know?

There is an estimated risk that out of one-thousand angioplasty procedures performed, one to four due to various reasons could result in death. There is also up to a five percent chance that there will be emergency open heart intervention and that as many as three percent of these procedures could induce a heart attack.

Cardiac Enzymes

Cardiac enzymes, which are also referred to as "serum cardiac markers," are blood tests that are utilized in the diagnosis of a myocardial infarction (heart attack). Since these tests are performed on the blood their results are considered to be more definite indicators, unlike the electrocardiogram (EKG) which can possess elevated ST segment (Q-wave MI) or non ST segment elevation (non Q-wave MI). As the heart suffers an infarction cardiac cells die through a process called necrosis. When the cells die, their membrane ruptures and substances called creatine kinase, myoglobins, isoforms and troponin are released into the bloodstream. Creatine are amino acids that are present in muscle tissue. Myoglobins are iron containing proteins that are present in muscle tissue that act as a storage basin for oxygen and carbon dioxide. These proteins are released into the bloodstream when cardiac muscle injury occurs. Kinase is a substance that is capable of activating zymogens. Zymogens are passive enzymes that can become active by interaction with kinase. Troponins are isoform proteins that do not normally present in the blood of healthy individuals. The troponin is considered the more specific test when evaluating for a heart attack. By measuring the levels of these substances within the blood it can indicate if a heart attack is occurring and by how high their levels increase, will provide information as to its intensity. Cardiac enzymes are blood tests that are performed to evaluate these levels. When patients present to a hospital or Physician's office with angina (chest pain/pressure) with complaints of accompanying symptoms such as nausea, dizziness, shortness of breath or sweating, cardiac enzymes blood tests are drawn and examined together with other tests such as the electrocardiogram (EKG) and vital signs. The serial cardiac enzyme test combines three blood tests drawn every 8 hours over a 24 hour period, they are referred to as the CPK (creatine phosphokinase), CK MB (CK fraction) and troponin. It is necessary however, to evaluate all three of these enzymes together to achieve a true result.

Normal Cardiac Enzyme Levels

CK 1 Men 25-130 micrograms per liter
Women 10-150 micrograms per liter
CK 11 up to 7 micrograms per liter
Troponin up to 0.4 nanograms per milliliter

* The cardiac troponin can rise to 40 times its normal level. This test will rise in 2-6 hours and will peak in 12-16 hours.
* The CPK will start to rise in 4-6 hours and peak in 24 hours. This substance is elevated in 90% of all MI's.
* The CK MB is a small fraction of the CPK and will rise in 3-4 hours. It will rise and return to its normal level sooner that the CK.
  Myoglobin rises in 2 hours and will peak in 6-8 hours. This test is now utilized less frequently.

Cardiac Ischemia

Ischemia is defined as the insufficient supply of oxygen-rich blood flow to the body's organs. This reduction or stoppage in blood flow can be precipitated by different factors such as a blockage caused by an embolism or thrombus, or it may be the result of a cardiovascular disease that has evolved over a period of years called Coronary Artery Disease. This is a disease that over time involves the accumulation and build-up of plaque and fatty deposits within the blood vessel that narrow the lumens and reduce the amount of blood delivered to the body. This can result in irreparable damage or death to the tissue and organs fed by the compromised blood vessel. Cardiac ischemia is a condition where there is a blockage in a coronary artery that prevents part of the heart muscle that is fed by a particular artery from receiving enough oxygen-rich blood. This lack of oxygen-rich blood can damage, weaken and cause death to the myocardial tissue. It can also affect the function of the conduction system and debilitate the hearts ability to perfuse a sufficient amount of blood to the rest of the body. Eventually this overload on the heart muscle leads to heart failure and creates damage to the rest of the body's organs (mainly the kidneys and brain) through necrosis or cell death. The effects and symptoms caused by the decreased blood supply brought on by this disease are more noticeable in times of exertion, exercise, or stress, when the hearts demand for more oxygen increases. It is during these episodes that an individual may experience short episodes of chest discomfort in the form of pain or a squeezing sensation, a symptom referred at as angina. Other symptoms that may be related to ischemia can include palpitations, the feeling of being tired or fatigued and shortness of breath. However there can also be ischemia present without an individual experiencing the common symptoms, this is referred to as "Silent Ischemia." In cases involving silent ischemia a heart attack can occur without warning that can result in sudden cardiac death. There are a few different tests that physicians utilize to evaluate for conditions that may be favorable for or have the potential to induce ischemia. Some of these tests may include:

* echocardiogram – this is a test that utilizes high frequency sound waves and echo technology to provide a physician with information regarding the hearts muscular movements and blood flow by allowing him to visualize it on a viewing monitor while it is beating.
* electrocardiogram (EKG) – this test records the electrical activity as it travels through the heart. The EKG test is most useful in identifying this condition through rhythm changes that occur as a result of the problem.
* stress test – this procedure utilizes both the Echo and EKG technologies while the patient exercises that provides the physician with a view of the patients electrical as well as mechanical heart function before, during, and after a stressful situation.
* pharmacological stress test – in this procedure the heart is chemically stressed using a drug called dobutamine while the physician observes and evaluates muscular reaction and blood flow on a viewing monitor.
* blood tests – monitoring levels such as cholesterol will allow for lifestyle changes to be initiated.

* rapid heart rates that limit the filling ability of the ventricles by not allowing enough time for them to refill.
* restriction or blockages within arteries caused by blood clots / embolisms
* hypotension – low blood pressure
* constrictive pressure on blood vessel such as that created by cardiac tamponade or other outside source

note:
ischemia – a shortage of blood being supplied to the body's tissue and organs.
hypoxia – a shortage of oxygen being supplied to the body's tissue and organs that may be caused by an insufficient amount of blood or a decreased amount of oxygen introduced into the bloodstream that is precipitated through diseases such as Congestive Obstructive Pulmonary Disease (COPD) that affects the transfer of carbon dioxide with oxygen.

Did You Know?

Your heart circulates your body's blood supply, which is about five quarts, through your body approximately one-thousand times per day. Each circulation takes approximately twenty-three seconds.

Every year four hundred thousand individuals die from complications related to cigarette smoking. Smoking can triple your chances of becoming afflicted and dying from heart disease at mid-life. Smokers have a twenty-two to one higher probability of contracting and dying from lung cancer than a non-smoker. The chance of contracting and dying from bronchitis and emphysema is increased by ten times.

Heparin Anticoagulant/Antithrombotic

Heparin is a true anticoagulant because it becomes part of (combines) with the blood to prevent or delay clotting. This substance is found in different body tissue, but mainly the liver, and is used as an agent to prevent coagulation and the formation of clotting in arrhythmias such as atrial fibrillation or conditions like deep vein thrombosis that can precipitate pulmonary emboli. It is also used as a anti-coagulation agent in procedures such as open heart surgery to inhibit clotting. Although heparin will not dissolve a clot that is already formed, this medication is used as a therapeutic measure in atrial fibrillation conditions such as chronic DVT to prevent clot formation due to the lack of blood volume movement.

Cardiac Tamponade

The pericardium is a thin sac made of fibrous tissue that contains about 25ml of fluid that surrounds and protects the heart. Cardiac Tamponade occurs when this sac becomes inflamed or there is an excess of fluid that has accumulated within it that causes there to be a constrictive pressure on the heart. Over time the pressure created restricts the hearts chambers from fully expanding during the refilling phase of the cardiac cycle and prevents them from taking a complete fill of blood. The inability to expand compromises their ability to produce enough pressure to perfuse blood to the lungs and body. This condition also compromises the blood flow to the hearts surface arteries, precipitating myocardial necrosis through ischemia. Cardiac Tamponade is classified in two different levels that are referred to as sub-acute or acute depending upon its severity and emergent status. Sub-acute allows a progressive accumulation of fluid that effused into the pericardium over a period of time. This gradual process gives the pericardium time to stretch during the effusion, allowing up to 1 liter or more of fluid volume to accumulate within the sac before the serious side effects of Tamponade occur. The other form of this condition is referred to as acute Tamponade and this form occurs when there is a rapid accumulation of excess fluid over a short time period that provides no time for the sac to stretch to accommodate for the increased volume. Acute Tamponade is a more serious situation that requires emergent steps to be employed to reduce the pressure created by the rapid fluid build-up. Pericardial fluid build-up can be idiopathic or it may be secondary to and the result of various medical factors such as cancer, viral infection, kidney failure, damage caused by thoracic surgery which is also referred to as iatrogenic trauma, chest trauma that is of a blunt nature or puncture wounds caused by stabbing or gunshots. This condition displays various symptoms such as weakness, fatigue, chest pain, shortness of breath, abdominal swelling, painful breathing, vein distention in the neck, and a bluish skin color resulting from the lack of oxygen-rich blood supplied to the surface arteries. Diagnosis of this condition is made with the utilization of both instrumentation and physical evaluation. The electrocardiogram or EKG displays changes such as elevated ST segment, cardiac cycle to cardiac cycle voltage changes of the QRS and T-wave and a rapid heart rate which is referred to as tachycardia. Tests such as the chest x-ray will show changes in the size and configuration of the heart, with the pressure applied to the heart causing it to present with a globular appearance. The echocardiogram presents an enlarged pericardium and in some cases collapsed ventricles due to the influence of the pressure. Hypotension caused by the decreased stroke volume precipitates low pressure venous blood return that presents itself through neck vein distension. Surgical treatment for this condition would center on the removal of fluid from the pericardium to reduce the pressure, then to repair the sites that are responsible for the effusion. One such procedure is called the pericardiocentesis. This procedure utilizes a needle that is inserted into the pericardium to withdraw the excess fluid. Another method is called the pericardial window which involves the surgical opening of the pericardium to allow excess fluid drainage.

Where's the Impulse?

The following descriptions will help in understanding where in the conduction cycle the electrical impulse block occurs. By observing the cardiac cycle it will help to determine its location.

Sinus Pause is a delay in the formation of the electrical impulse at the site of the sinoatrial node. This can be caused by disease tissue within the sinoatrial node, or it can be precipitated by certain Class 1 # 3 antiarrhythmia medications. This conduction anomaly will produce an extended time delay between complete cardiac cycles (PQRST) that can measure up to 3 seconds before it becomes classified as a cardiac arrest.

Sinus Arrest occurs when there is a failure at the site of the sinoatrial node to create an electrical impulse for a period greater than 3 seconds. As with the sinus pause there is prolonged time duration between complete cardiac cycles (PQRST). If the automaticity within the AV Junction or the Purkinje fibers does not take control of the conduction of the heart, ventricular standstill will occur and death will follow.

Slow Ventricular Response (SVR) is a conduction abnormality that is directly related to the atrial fibrillation arrhythmia. The extended time durations precipitated between ventricular cycles by this arrhythmia is an electrical impulse regulating issue as it is being transmitted from the atria to the ventricles. In this instance there is erratic atrial electrical activity present that bombards the AV node causing it to deliver electrical impulses in an irregular fashion. At times the impulse transmission to the ventricles will slow to a level that it presents longer extended time durations between the ventricular complexes which slows the heart rate. Treatment would be to restore sinus rhythm.

Sinus Exit Block is a problem that occurs when the electrical impulse is blocked after the sinoatrial node gererates it and it is being delivered to the atria. As with the atrioventricular nodal blocks there are various degrees that are classified in three different categories. Type 1 will display a progressively shortening PR interval until there is a missed cardiac cycle (PQRST). Type 2 will display missed cardiac cycles that occur suddenly and without warning that will measure out to be equal to that of the normal cardiac cycles. Type 3 is hard to distinguish from sinus arrest and will be comprised of multiple missed complete cardiac cycles.

AV Node Blocks differ from the SA Nodal blocks in the location within the cardiac cycle where the block occurs. The difference can be seen on an EKG tracing by observing the configuration of the P-wave and QRS complex. In SA blocks the PQRST cycle is complete due to the fact that the block took place before it could stimulate the atria, causing it to skip a complete cardiac cycle. With the AV blocks the P wave and QRST are separated. This indicates that the block occurred further along the conduction pathway as the impulse was being transmitted from the atria to the ventricles.

Bundle Branch Block results from damage to the right and left bundle branches or fascicle fibers. These are located at a point past the AV junction where the Bundle of His separates into two distinct branches.

First Degree AV Block occurs when diseased tissue or certain Class 1 and 3 antiarrhythmia drugs impede the electrical impulse as it passes through the AV junction. This conduction abnormality is considered more of a conduction delay as opposed to an actual heart block due to the fact that all electrical impulses eventually reach the ventricles.

Cardiomyopathy

Cardiomyopathy is a disease that inhibits the heart from performing its normal function. This condition can be primary or secondary and is classified as ischemic or non-ischemic. Non-ischemic has three different variations that are hypertrophic, dilated, and restrictive. Ischemic cardiomyopathy is directly related to a diminished supply or lack of blood to the heart muscle. Primary causes would be considered to be those that can't be the result any specific origin. Secondary is linked to an origin that presents itself to be the cause that perpetuated the condition such as coronary artery disease or hypertension. These factors affect the myocardium by causing the muscle fibers to become enlarged and less flexible, resulting in weak contractions that reduce its effectiveness as a pump and diminishes blood movement throughout the body. This will eventually lead to a condition called congestive heart failure. One of the major complications of this condition is the formation of blood clots due to poor blood circulation which causes blood to pool within the hearts chambers from lack of movement. These clots could then enter into the circulatory system, presenting a situation that is favorable for a stroke or heart attack. It is estimated by the American Medical Association that cardiomyopathy is responsible for roughly twenty-seven thousand deaths per year. Although there are different variables of this condition, they are separated into two groups. The two variations are:

* Ischemic Cardiomyopathy – one cause of this form of cardiomyopathy is coronary artery disease. This disease affects the coronary arteries that feed the heart oxygen-rich blood. The diminished blood flow caused by the narrowed arterial passageways related to this disease starve the heart muscle of the necessary oxygen supplied by the blood, causing myocardial damage and adversely affecting the function of the heart muscle.
* Non-schemic cardiomyopathy - non-ischemic cardiomyopathy has three variations that are caused by different complications, they would include:
  * Hypertrophic Cardiomyopathy - which is a condition in which the heart muscle is thickened due to abnormal growth of the muscular fibers. It has been shown through research that individuals with hypertrophic cardiomyopathy can possess eight times the amount of connective tissue of that of a person not afflicted with this condition. With this added mass, the muscle is not allowed to rest during the repolarization phase of the cardiac cycle.
  * Dilated Cardiomyopathy – this is a situation where the ventricles become enlarged and weakened. The cause of this type can sometimes be unknown, but it is known to occur when factors such as alcohol abuse, hypertension, genetics, or viral infection are present. It is also suspected that this form of cardiomyopathy may be the result of the cardiac tissue being attacked by the immune system which mistakenly sees it as a foreign body.
  * Restrictive Cardiomyopathy – restrictive in it's inability to refill with a complete supply of blood during the repolarization phase. This form does not normally display enlarged ventricles, and is caused by an inflammation that is precipitated by white blood cells.

Treatments for this condition could include the following measures:

* Anti-coagulation therapy to reduce the possibility of blood clots . In this treatment drugs such as heparin or coumadin are administered to aid in the prevention of clot formation
* Minimally invasive procedures such as angioplasty would increase the blood flow by opening up the coronary arteries and increasing the oxygen-rich blood supply to the heart.
* The use of medications that would lower the blood pressure and lessen the load on the heart. This is accomplished with the use of ACE inhibitors and beta-blockers. This therapy would not only reduce the hearts workload, but it would help reduce the possibility of arrhythmias.

Did You Know?

The heart begins to develop as soon as the embryo starts developing inside the mothers womb, with the chamber walls taking form as soon as 6 weeks old. It grows so fast that there is not enough room for it to become longer, so it doubles back and twists, taking on the appearance we recognize. There is a layer of tissue called the septum that grows down the middle at this point dividing the right and left sides. A male heart weighs approximately 11 ounces, compared to 9 ounces of the heart of a woman. It is thicker at the top, is approximately five inches long, three & one-half inches wide, and two & one-half inches thick. It is located in the middle of the chest in a horizontal position with the lower part of the heart pointed to the left side of the body. The left ventricle muscle walls are three times as thick as the right to provide a better catalyst to move the blood throughout the entire body. It beats 70 times per minute (normal heart rate), or 100,800 times a day, 90 times a minute for a 7-year-old child, and 120 times a minute for an infant. The heart pumps five quarts of blood through its chambers every 60 seconds.

A thin sac that is called the pericardium surrounds the heart. This protective envelope is made of a tough tissue that acts as a buffer to protect the heart from damage caused by contact between itself, and the lungs and chest wall. There is a fluid that is discharged between it's smooth interior wall lining and the heart to act as a lubricant to protect it from friction caused by rubbing.

Death that results from cardiac arrest is referred to as "Sudden Cardiac Death". It is estimated that there are as many as three-hundred thousand occurring each year, with the probability that most of these situations could have been reversed if prompt medical intervention could have been obtained.

Cardioversion

The cardioversion procedure employs the intentional introduction of an electrical shock to the heart through the chest wall to induce the momentary depolarization of the myocardium. This interruption presents the opportunity for the sinoatrial node to resume control as the hearts dominant pacemaker of the conduction system. This procedure is performed to convert narrow complex supra-ventricular arrhythmias such as rapid atrial fibrillation and atrial flutter back to normal sinus rhythm and is also used to terminate life threatening wide complex arrhythmias that present with a pulse such as ventricular tachycardia and torsades de pointes. The cardioversion can be performed by utilizing a couple of different methods that involve the use of both electric shock and cardiac drugs. One such method is the use of an external defibrillator equipped with EKG rhythm display screen similar to the type utilized on code carts used for emergencies occurring on cardiac units in hospitals. In this scheduled procedure which is done in a controlled environment, a sedative is given to the patient and pads with electrodes that are connected to the defibrillator unit are placed on the chest and back. At the point in the cardiac cycle that is timed to occur simultaneously with the R-wave, an electrical shock that is pre-set in current and time duration is delivered. A shock that is not delivered in the proper moment of the cardiac cycle that falls into the relative refractory period of the ventricular repolarization can induce ventricular fibrillation. The next form of cardioversion is used in emergency situations by EMT's and other emergency personnel utilizing a portable external defibrillator or AED which stands for "Automated External Defibrillator." In this process the electrical shocks are administered by EMT personnel by pressing a button at the prompting of the AED unit which contains a microprocessor that evaluates the patient's heart rhythm. This form of cardioversion is utilized during emergency situations that involve life threatening arrhythmias such as ventricular tachycardia. The ICD or "Implantable Cardioverter Defibrillator" is a form of cardioversion that incorporates a device that is implanted into the chest of the individual. This unit acts as a permanent on-board cardioversion unit that continuously monitors the hearts rhythm, automatically delivering an electric shock to terminate arrhythmias such as ventricular tachycardia and rapid atrial fibrillation as thy appear. These devices are used in patients with chronic arrhythmia recurrences that are uncorrectable through the use of other means. This last method involves the use of medications and is referred to as "Chemical Cardioversion" or "Pharmacologic Cardioversion." In this type of cardioversion, antidysrhythmic medications are administered to correct the arrhythmia by manipulating the electric impulse at specific transmission sites in the conduction system. Some of these medications could include antidysrhythmics such as adenosine which induces heart block, or amiodarone, cardizem, or rhythmol that stabilize the heart by controlling the heart rate. After cardioversion, normal sinus rhythm is maintained through the use of medication taken in maintenance doses by the individual.

Note: Defibrillation is a term that is used to describe a procedure that is used in emergency situations on arrhythmias such as ventricular tachycardia and ventricular fibrillation that present without a pulse. Defibrillation of an arrhythmia that presents with a pulse is referred to as cardioversion.

Chronic Obstructive Pulmonary Disease

Also called "COPD," is a progressive recurring respiratory disease that compromises the body's ability to provide an adequate amount of air intake or effectively exchange carbon dioxide with oxygen within the lungs. This disease affects roughly twenty percent of the adult population, generating over thirteen million Physician office visits annually. It is also the fourth leading cause of death in the United States. Chronic Obstructive Pulmonary Disease is enhanced by cigarette smoking which can claim some degree of responsibility in over eighty-five percent of reported cases. The primary symptoms related to this disease would include shortness of breath, wheezing and coughing. There are two main diseases that are associated with chronic obstructive pulmonary disease; they are chronic bronchitis and emphysema. Chronic bronchitis is classified as a sputum producing cough in the presence of obstructed air flow for three months within a two year period. This is a condition where there is the presence of inflammation of the bronchi or airways of the lungs that restricts the amount of air intake. This situation decreases the amount of oxygen in the blood, creating a condition that is referred to as hypoxia. Severe drops in surface arterial oxygen saturation levels such as 85% or less can cause the skin to present with a bluish tint, a condition called cyanosis. Bronchitis is diagnosed by using various methods such as a chest x-ray that will reveal hyperinflation of the lungs, blood tests such as C-reactive protein (CRP), white blood cell count also referred to as WBC that will present with elevated levels, the presence of a persistent cough and abnormal lung sounds. Emphysema is a problem that is associated with the air sacs within the lungs exchanging newly inhaled oxygen with carbon dioxide that is the waste product air from which the oxygen has been extracted by the body. The main cause of this disease is the inhalation of smoke from years of tobacco use. This activity causes lung tissue to lose its flexibility, the capillaries to deteriorate and the smaller airways to collapse. This leaves un-exhaled air within the alveoli causing them to remain expanded which precipitates them to rupture. With emphysema, unlike bronchitis, it is possible to maintain adequate oxygen saturation levels with the blood by hyperventilating until the disease deteriorates to a degree that hyperventilation no longer compensates. Emphysema can be diagnosed utilizing pulmonary function tests also referred to as PFT's, chest x-rays and blood tests. Complications associated with chronic obstructive pulmonary disease make carrying out routine daily activities harder or not possible at all.

Did You Know?

Every year four hundred thousand individuals die from complications related to cigarette smoking. Smoking can triple your chances of becoming afflicted with and dying from heart disease in mid-life. Smokers have a twenty to one greater probability of contracting and dying from lung cancer than a non-smoker, with an increase of ten times that they will contract and die from bronchitis or emphysema.

Congestive Heart Failure

Congestive Heart Failure, also referred to as "CHF," occurs when the heart loses its ability to effectively pump oxygen-rich blood to the body and lungs. This condition not only precipitates fluid build-up within the body's tissue and lungs, but also has the potential to cause irreversible damage to the heart muscle. The lack of oxygen-rich blood that is supplied to the body requires the heart to work harder to satisfy its needs, precipitating conditions such as hypertrophy where the heart becomes enlarged and loses it flexibility. There are different factors that may be responsible for this condition. Some of these would include:

* less contraction strength due to reasons such as a heart attack or infection
* compromised blood flow volume precipitated by defective valves
* arrhythmias / irregular heartbeats
* untreated continuously high blood pressure
* diseases such as Coronary Artery Disease put additional stress on the heart muscle by making it work harder to provide an adequate amount of blood through narrower blood vessels

This condition presents itself with various symptoms that may include an increased heart rate, chest pain, fluid weight gain and the inability to perform daily routines due to fatigue. There may also be secondary symptoms throughout the body that are directly related to the function of each individual side of the heart, the are:

* left-sided heart failure – failure of the left ventricle weakens the contraction strength resulting in less drive during perfusion, causing a build-up of blood returning to the heart within the veins. The fluid portion within the veins then enters into the surrounding tissue precipitating a condition referred to as edema. Edema can then spread to various parts of the body, normally within the tissue of the extremities furthest from the heart muscle such as the feet and ankles. However this condition can be present within the stomach and liver as well. Left sided heart failure can also put unmanageable stress on the kidneys by compromising their ability to properly dispose of excess fluids containing certain elements such as salt, creating a situation that would eventually result in kidney failure.
* right-sided heart failure – this side of the heart feeds the lungs waste blood that needs to be re-oxygenated. Abnormal function of the right ventricle would result in fluid build-up within the lungs causing a situation where the individual would experience shortness of breath created by the body's inability to take in enough air. The result would be an inadequate amount of oxygen to re-oxygenate the waste blood. As a result the individual would experience fatigue, and it would become probable that they would develop a cough.

Treatment for Congestive Heart Failure can be administered in a few different ways, but lifestyle changes are necessary to reap the benefits. The first method of treatment would involve medications to perform different tasks, a few of these would include:

* diuretics such as lasix to remove fluids from the body and lungs. This drug causes the individual to urinate more frequently to remove excess fluid, which can also decrease the chemical elements within the body such as potassium and magnesium. To avoid adverse affects to the body's organs caused by levels of these elements dropping too low, routine blood testing is done and if necessary replacement supplements are re-introduced into the system.
* cardiac medications such as digoxin (digitalis) are administered to enhance and strengthen the contractions of the heart. This provides more catalyst to deliver blood throughout the body during the left ventricular depolarization (contraction) phase of the cardiac cycle.
* beta-blockers are used to help control blood pressure and lower the heart rate to a normal level. One example of a beta-blocker that might be used in this situation would be the drug called lopressor.
* blood thinners or anti-coagulants such as heparin and coumadin are used to aid in the prevention of clotting caused by the pooling of blood due to poor circulation. Although these drugs cannot dissolve clots that are already present, these drugs possess properties that prevent future clots from forming.
* a drug referred to as Natrecor is given in IV Therapy to help the heart. This medication is produced to mimic the body's hormone "Brain Natriuretic Peptide," also called "B-type Natriuretic Peptide," or "BNP," which is naturally secreted by the body to help in times that it is experiencing too excessive a workload.

There is also the option that involves the utilization of a bi-ventricular pacemaker, a process that is referred to as "resynchronization therapy." As the conduction system ages it sometimes loses its ability to deliver a synchronized electrical impulse to both ventricles. This causes the walls of the left ventricle to contract out of synchronization due to excessive flexibility of the septum, resulting in less force behind the left ventricular chambers perfusion of blood to the body. Resynchronization therapy corrects this abnormality by utilizing a pacemaker that that is designed to deliver a programmed electrical impulse to both ventricles simultaneously to induce synchronized contractions. This causes the septum between the ventricles to be firm from the pressure of the right ventricular depolarization, increasing the force of the blood delivery from the left ventricle to the body.

Did You Know?

The first heart transplant was performed in the year 1967 by a surgical team in Cape Town, South Africa. The surgery was initiated and headed by a surgeon named Christian Barnard. Most of the heart transplants in the early stages of research and experimentation resulted in the deaths of the recipients within the first year. In the 1980's Stanford University through technological advances and the introduction of medications to promote the body's acceptance of the new organ, raised the survival rate for the patients to live past the first year to over seventy percent.

Coronary Angiogram

A Coronary Angiogram is an exploratory procedure that is performed by a Cardiologist to evaluate the coronary arteries located on the surface of the heart. These arteries provide the heart muscle with oxygen-rich blood that possesses the nutrients needed to ensure its proper function. There are over one million cardiac angiograms performed yearly that are ordered following abnormal results from tests such as the echocardiogram, stress test (exercise or chemical), or the EKG (electrocardiogram). These procedures may also be ordered when there is the presence of angina, shortness of breath, heart arrhythmias, or other symptoms that may be an indication of restricted blood flow. Over time fatty deposits and plaque accumulate within the blood vessels creating a condition referred to as Coronary Artery Disease. This condition causes a constriction in the arterial pathway (lumens) that lessens the volume of oxygen-rich blood that can be carried through them. This diminished blood supply creates a situation where the heart is starved of the nutrients it needs which can result in episodes of angina or in severe cases a heart attack (myocardial infarction). The coronary angiogram procedure uses a contrast dye (contrast medium) in conjunction with x-ray technology to provide a visible image of the coronary arteries as the blood circulates through them. To perform this test a cut is made in a blood vessel as an entry site for insertion of the guide wire and catheter tube. The most common location is the femoral artery, although in certain instances the arm is used as an entry site. The thin catheter tube is then inserted through the incision along the guide wire and a contrast dye is injected through the tube into the circulatory system. With the use of the x-ray the physician can observe the dye as it travels through the circulatory system, which allows them to locate compromised areas that have been narrowed and to what extent. This procedure is employed during a cardiac catherization during information gathering to form a diagnosis and to explore the available treatment options such as the utilization of medications, the PTCA (Precutaneous Transluminal Coronary Angioplasty), or the CABG (Coronary Artery Bypass Graft). Although this procedure is minimally invasive and complications are considered low, less than 3%; the procedure does possess risks. A few of these risks would include damage to blood vessels, allergic reaction to the contrast dye, triggered onset of cardiac arrhythmias, heart attacks, or strokes.

Did You Know?

There is an estimated risk that out of every one thousand angioplasty procedures performed that one to four due to various reasons could result in death, there is up to a five percent chance that there could be emergency open heart intervention, and up to a three percent chance that the procedure can induce a heart attack.

Pounds to Kilograms Conversion Table

Pounds	Kilograms	Pounds	Kilograms
100	45	205	92.25
110	49.5	210	94.50
115	51.75	215	96.75
120	54	220	99
125	56.25	225	101.25
130	58.50	230	103.50
135	60.75	235	105.75
140	63	240	108
145	65.25	245	110.25
150	67.5	250	112.50
155	69.75	255	114.75
160	72	260	117
165	74.25	265	119.25
170	76.50	270	121.50
175	78.75	275	123.75
180	81	280	126
185	83.25	285	128.25
190	85.50	290	130.50
195	87.75	295	132.75
200	90	300	135

Formula used: lbs x 0.45 = kilograms

Did You Know?

The conduction phenomenon that is referred to as "Wenckebach," "2nd Degree Atrioventricular Block type 1," or "Mobitz 1," was first documented in the year 1899 by Karel Frederik Wenckeback in a paper that he had written titled "On the Analysis of Irregular Pulses." In this publication he explained how the blockage of electrical impulses in the atrioventricular node of frogs caused the PR interval to lengthen which resulted in the intermittent loss of a ventricular conduction.

Coronary Artery Disease

Coronary Artery Disease, which is also referred to as Atherosclerosis or CAD, is the accumulation of plaque on the interior walls of the coronary arteries that feed oxygen and nutrient rich blood to the heart. This build-up causes the loss of elasticity in the artery walls and constricts the opening of the passageway which causes a reduction in the hearts supply of oxygen-rich blood. This creates a condition called cardiac ischemia which occurs when there is an insufficient amount of oxygen that is being supplied to the tissue to satisfy its need. As a result the individual experiences angina (chest pain) that depending on the level of severity of the restriction of blood flow can occur both during exertion which is referred to as stable angina or can be the more serious form that becomes more intense and can occur at rest. This is called unstable angina and is a warning sign that the heart muscle is not receiving enough oxygen-rich blood when the heart is under minimal strain and without much more of an increase other than its normal workload. Symptoms of angina that present at this stage (unstable) can result in a heart arrhythmia or a heart attack (myocardial infarction) which kills myocardial cells and tissue and incapacitates the heart muscle's ability to function properly. This process is called necrosis and once it occurs is irreversible. Coronary artery disease is suspect if there are complaints of chest pain that may radiate to the back, left arm, neck and jaw, shortness of breath, fatigue, abnormal EKG changes such as ST depression that occurs during a stress echocardiogram within the exercise period of the test (as compared to a resting EKG taken before the stress test). When coronary artery disease is diagnosed its severity is determined through the use of an intravascular ultrasound, MRI, or coronary angiogram that is performed during a cardiac catherization procedure. During the coronary angiogram test, a contrast dye is introduced into the cardiovascular system through a catheter tube and allowed to flow through the coronary arteries. X-rays are then taken that show the locations and severity of any restrictions within the coronary artery passageways. At this point there are a few options that may be utilized depending upon the severity of the blockages that can involve medicinal intervention with drugs such as nitroglycerine, beta blockers and drugs that counteract cholesterol to a medical procedure referred to as the coronary artery bypass graft (CABG). However, the treatment that is utilized most is a minimally invasive procedure referred to as the Percutaneous Transluminal Coronary Angioplasty (PTCA) which is commonly called the angioplasty. This involves the employment of an inflatable balloon that is fed through a catheter into an artery (commonly the femoral artery located at the site of the groin) to the location in the coronary arteries that presented with the blockage. When it arrives at the compromised location in the artery the balloon is inflated and the plaque is compressed back against the artery wall, opening up the passageway to provide increased blood flow. During this procedure it is also sometimes necessary to deploy an expandable wire tube called a stent to provide structural integrity to weak vessel walls. This PTCA process is also referred to as "Percutaneous Coronary Intervention" or "Reperfusion Therapy." When the angioplasty fails to produce acceptable results a more invasive surgical procedure called the "Coronary Artery Bypass Graft" or "CABG" is performed.

The CABG procedure requires the opening of the chest to expose the heart to allow the surgical re-routing of blood flow around the blockage with blood vessels commonly harvested from the legs. Although coronary artery disease may take years to progress to the level that it presents noticeable symptoms, it is the number one cause of heart disease and death in the United States. It is also said to be one of the most common reasons for premature death among those greater than 20 years of age. Family history, age and gender are non-alterable circumstances that put one more at risk for developing coronary artery disease, however there are conditions that contribute to CAD that can be controlled to lessen the possibility of acquiring the disease. A couple of these factors would include conditions and diseases such as hypertension and diabetes which can be controlled with drug intervention, or cigarette smoking, obesity, sedentary lifestyle, hypercholesterolemia and diet which can be corrected with lifestyle modifications.

Did You Know?

* Incidents that involved the restenosis (re-narrowing) of blood vessels that have been re-opened through a procedure called angioplasty was decreased due to new medicated stents named "Cyphers." These stents were developed by Johnson and Johnson and approved for use by the Food and Drug Administration in the year 2003. They are designed to release a special anti-rejection medication during the healing process that inhibits excessive cell growth.
* There are one and one-half million heart attacks annually in the U.S. with a resulting five- hundred thousand deaths.
* Every year four hundred thousand individuals die from complications related to cigarette smoking. Smoking can triple your chances of becoming afflicted and dying from heart disease at mid-life. Smokers have a twenty-two to one higher probability of contracting and dying from lung cancer than a non-smoker. The chance of contracting and dying from bronchitis and emphysema is increased by ten times.
* Statistics show that women have a greater risk of dying from some form of cardiac related disease. In the year 2003 there were over two-hundred thirty thousand women who died from cardiovascular disease. Over eighty thousand died from a heart attack with an average age of seventy years old. Of the heart attack survivors there were more than thirty five percent that expired within the first year of the initial onset. Women on average take up to two hours longer than that of a man to obtain an examination when having heart attack symptoms, reducing the window of opportunity to benefit from therapy employed to dissolve clots. There is also a fifty percent higher mortality rate from heart attacks for women under fifty years of age as opposed to males within the same age group
* Being overweight increases the risk of becoming afflicted with heart disease. It is estimated that sixty percent of the adults in the United States are obese and that forty percent engage in minimal or no physical activity. Studies have also shown that obesity in children has almost tripled in the past fifteen years.

CABG.

CABG, which stands for coronary artery bypass graft, is a surgical procedure that is performed to revascularize the myocardium to reduce the risk of a myocardial infarction (heart attack) and eliminate chest pain (angina) caused by ischemia. Over the years plaque accumulates within the arterial lumens and decreases the oxygen-rich blood flow that is vital to the survival of the body and its organs. The reduction or complete stoppage of blood flow causes the body's cells to die within a very short time. In this instance the cardiac cells of the heart muscle. The coronary artery bypass graft is a procedure that is performed after less invasive options to revascularize the myocardium have failed. These options could include drug therapy, or the PTCA (percutaneous transluminal coronary angioplasty), which is a minimally invasive procedure that utilizes an inflatable balloon to push plaque back against arterial walls to reopen the lumens (passageways). These blocked sections of untreatable coronary artery are then bypassed with pieces of vein or artery harvested from other locations such as the saphenous veins within the legs or the radial arteries taken from the arms. The bypass procedure can include the single bypass (one coronary artery bypassed), the double bypass (two coronary arteries bypassed), triple bypass (three coronary arteries bypassed), quadruple bypass (four coronary arteries bypassed), quintuple bypass (five coronary arteries bypassed). This decision regarding which vessels and the number of affected vessels to be repaired is determined by the results of a coronary angiogram study that is performed during the cardiac catherization procedure and observations made by the physician after the heart and its arteries are visible for evaluation. A few of the more common arterial locations this surgery would be performed on would include the RCA (right coronary artery), LCA (left coronary artery), LAD (left anterior descending artery) and LCX (left circumflex artery). This surgical procedure can be performed by either one of two different methods. The standard method which is referred to as "cardiopulmonary bypass" or "on-pump coronary artery bypass" was first successfully performed in the year 1953 by a cardiac surgeon and inventor of the heart lung machine in Philadelphia named Dr. John Gibbon to repair an ASD (atrial septal defect) in an 18 year old girl named Cecelia Bavolek. This procedure utilized the newly developed heart lung machine and took approximately 45 minutes to an hour with the patient fully dependant on the heart lung machine for approximately 25 minutes. This procedure is employed due to the difficulties involved in operating on a moving heart. To accomplish this it is necessary to temporarily stop the heart using a chemical that induces cardioplagia and protects the myocardial tissue from damage during the stoppage. After the cardioplagia is completed the function of the patient's heart and lungs are replaced by a heart lung machine which removes deoxygenated blood, replaces it with oxygenated blood and mechanically circulates it through the body. Without perfusion to the body's tissue and organs they begin to suffer permanent damage within four minutes. The second method that can be employed is called "off-pump coronary bypass surgery" or "beating heart surgery" and has been in practice since 1985. This procedure does not incorporate the heart lung machine but uses myocardial "stabilizers" to neutralize the area of the heart that is being worked on by the surgeon.

Although this is a relatively new procedure and cannot be used for surgeries such as valve or atrial septal defect (ASD) repair, it constitutes about 25% of all coronary artery bypass surgeries performed and is a necessary form of heart surgery where stoppage of the heart would cause more damage or when the patient presents with other diseases. As with the standard or on-pump CABG, the chest is opened and the sternum (breast bone) is separated and pulled back with retractors to expose the heart and its arteries. The "target" arteries are then bypassed by grafting harvested vessels to a healthy artery or a point before the blockage to a location beyond the blockage, diverting the blood flow around the blockage. To perform this operation, extensive training is required by the surgery team to ensure adequate myocardial muscle perfusion is maintained during the procedure.

Complications from this type of surgery can be either minor or more serious in nature and can include infection, bleeding, heart attack, stroke, pneumonia and kidney failure. However, it has been documented that off-pump coronary artery bypass surgery reduces surgical wound infections by as much as 48% and the instances of surgery induced heart arrhythmias such as atrial fibrillation by as much as 30%. It has also been shown to present with fewer complications normally associated with on-pump coronary artery bypass and to have a shorter patient recovery period.

Did You Know?

The heart begins to develop as soon as the embryo starts developing inside the mother's womb, with the chamber walls taking form as soon as 6 weeks old. It grows so fast that there is not enough room for it to become longer, so it doubles back and twists, taking on the appearance we recognize. There is a layer of tissue called the septum that grows down the middle at this point dividing the right and left sides. A male heart weighs approximately 11 ounces compared to 9 ounces of the heart of a woman. It is thicker at the top, is approximately five inches long, three & one-half inches wide and two & one-half inches thick. It is located in the middle of the chest in a horizontal position with the lower part of the heart pointed to the left side of the body. The left ventricle muscle walls are three times as thick as the right to provide a better catalyst to move the blood throughout the entire body. It beats 70 times per minute (normal adult heart rate) or 100,800 times a day, 90 times a minute for a 7-year-old child, and 120 times a minute for an infant. The heart pumps five quarts of blood through its chambers every 60 seconds.

It is estimated by the FDA that in the year 2006, 25% percent of the heart attacks (roughly around one-hundred and seventy-five thousand) were going to be "silent." These events will display no symptoms, but unfortunately, heart attacks evolve causing myocardial damage over time and these unrealized cases take more time to receive medical attention.

Defibrillation

Defibrillation is a process by which an electrical shock is intentionally introduced to the heart to act as a counter measure in an attempt to correct arrhythmias such as ventricular fibrillation and ventricular tachycardia. The purpose is to stop all electrical activity (depolarize) in the hearts conduction system in an attempt to allow it to return it to normal sinus rhythm when it re-starts. Defibrillation was invented in the year 1899 by two Italian Physiologists named Dr.Prevost and Dr. Batelli who found that an electrical shock that was introduced into a ventricular tachycardia arrhythmia in a dog converted the arrhythmia back to a normal sinus rhythm. Although this breakthrough was established in the 1890's, the first human life saved by defibrillation did not occur until the year 1947 when Claude Beck was successful in using defibrillation to revive a patient. The use of the defibrillator with mobile capabilities to be carried by emergency personnel such as EMT's on duty was developed in the year 1966 by a cardiologist from Ireland named Dr. Frank Pantridge. This important development made life saving possible outside of a hospital and at the scene of the medical emergency.

The defibrillators used today come in various configurations that are designed to provide the compatibility to accommodate many different situations. Some of these different models would include:

Internal Defibrillator – this device is surgically implanted in a pouch just under the skin within the chest of an individual who suffers from recurrence of tachycardic arrhythmias. Implantable Cardioverter Defibrillators, also called ICD's, are programmed to sense for heart wave forms, heart rate and rhythm. These units are programmed to automatically deliver a shock if an arrhythmia is detected to terminate the rhythm.

Automated External Defibrillators (AED's) – the AED is carried and employed by emergency personnel such as EMT's and firefighters, but can also be found in places where there are large numbers of people at any given time. A few examples of where they might be found would be locations such as amusement parks, large sporting and musical events, casinos, schools and airports. This model is designed to be portable, possessing the capability of being delivered to the scene of the emergency. Its components consist of a battery, electrodes, and a control computer (microprocessor). When attached to the individual the microprocessor automatically assesses the rhythm of the patient and if necessary charges the unit and advises the emergency personnel that a shock is necessary. Although all other functions of this unit are automatic, delivery of the shock for defibrillation is still administered by emergency personnel through the activation of a button.

External Defibrillator – the external defibrillator is designed to be utilized in hospital settings in conjunction with a mobile emergency storage unit that is referred to as a code cart. This chest of drawers on wheels is stocked with equipment and medications to provide immediate emergency interventional measures during potentially fatal situations. The external defibrillator can be of the same functional design as the AED, or it can be a manual type that incorporates paddles that are hand-held at specific positions on the individuals torso to deliver electrical shocks implemented by a trained medical professional. These particular devises can also be utilized for performing cardioversions to convert uncontrollable tachycardic arrhythmias back to sinus, or to provide external pacing.

Ebstein's Anomaly

Ebstein's Anomaly, also called tricuspid valve displacement, is a rare congenital heart defect that was first recognized and described in the year 1866 by Wilhelm Ebstein. The abnormality involves the tricuspid valve which is a three flap valve located between the right atria and right ventricle that regulates the blood flow between chambers and forms a seal during the ventricular contraction. This anomaly randomly occurs in only 1 in 50,000 individuals, affects males and females equally, has a tendency to occur more in white people and when present in infants is referred to as "blue baby." In this abnormality the tricuspid valve is out of place or possesses deformed flaps causing there to be an inadequate seal, possibly caused by its flaps becoming stuck on the walls of the heart during the right ventricular chamber depolarization. The imperfect seal allows a back-flow of blood from the right ventricle into the right atria diminishing the amount of blood sent to the lungs for re-oxygenation. This is a condition that is referred to as tricuspid regurgitation. Serious cases of this anomaly can lead to further heart complicatations such as congestive heart failure, and is sometimes accompanied by atrial septal defect or "ASD" which is a hole that is located in the septum between the right and left atria. This can also precipitate a condition called cyanosis. This happens when the surface arteries receive an insufficient supply of oxygen which is a condition referred to as hypoxia, causing the skin to possess a bluish tint. Other symptoms precipitated by this anomaly may include rapid heart rates such as those seen in supraventricular tachycardia arrhythmias that can possess heart rates greater than 160 beats per minute, fatigue, syncope, fluid retention, lightheadedness and a decreased ability to perform daily functions. These symptoms can be an indicator that lifestyle changes, medications to control the heart rate, or surgical intervention to either repair or replace the defective valve may be required. This condition needs to be regularly monitored through different tests such as the echocardiogram which utilizes high frequency sound waves with echo technology to provide a real time image of the functioning heart. Another test that is utilized is the electrocardiogram, also referred to as the EKG. This test senses the hearts electrical impulse through electrodes placed at different positions on the limbs and torso. These signals are then transmitted to a receiving unit that amplifies the signals and provides a graphed tracing to be evaluated for conduction abnormalities. This test can also be performed using a Holter monitor which is a portable EKG vest that is worn by the individual for a prescribed amount of time to study and record cardiac electrical activity that occurs during normal daily activity.

Did You Know?

Death that results from cardiac arrest is referred to as "Sudden Cardiac Death." It is estimated that there are as many as three-hundred thousand occurring each year, with the probability that most of these situations could have been reversed if prompt medical intervention could have been obtained.

Echocardiogram

The echocardiogram, also called a "Transthoracic Echocardiogram" is a non-invasive test that uses high frequency sound waves and echo technology to produce the beating hearts image on a viewing screen. Sound waves are transmitted through a transducer that is moved over the chest area above the heart to form an image to be monitored on a video screen by the physician. Echocardiograms are employed by a Cardiologist to learn various types of information about the heart. Through the use of this echo technology, important information that otherwise would not be attainable is presented for evaluation. Some examples of what can be studied may include the hearts size and location, movement and function of its chambers and valves, the direction and velocity of blood flow, diagnostic information related to congenital defects, evaluation of the integrity and fluid content of the pericardial sac and to establish the presence of blood clots. There are a different types of echocardiograms that may be utilized depending on what information the Cardiologist is interested in studying. The most common tests performed would include:

* Stress Echocardiogram – the objective of this form of echocardiogram is to evaluate the hearts activity (particularly the wall motion) as it responds to exercise as compared to its activity at rest. This procedure is given in three phases:
  1) the patient is given an ultrasound while at rest and the activity is recorded to be compared to cardiac activity during later phases of the test.
  2) the patient is required to exercise on a treadmill to induce the heart to increase its activity for study under a stressful situation.
  3) a second ultrasound is given to the patient immediately after the completion of the exercise phase of the test to examine for changes in the motion of the myocardial walls or electrocardiogram (EKG) waveforms.
* Transesophageal Echocardiogram – also called a "TEE." This form of echocardiogram utilizes an endoscope that is inserted into the throat and through the esophagus. When the endoscope is positioned correctly sound waves are sent out to the heart and the reverberations are received to become an image to be displayed on a monitor.
* Dobutamine Stress Echocardiogram – the dobutamine stress echo is performed on patients who for various reasons are unable to utilize the treadmill. Dobutamine is a drug that enhances the hearts contractions that results in increased cardiac output. As with the other forms of echocardiograms this will aide in the forming of a diagnosis in cases where cardiac abnormalities that involve defective valves, heart failure, or myocardial defects are present. In this instance it is utilized to put stress on the heart to observe its function under drug induced conditions.

Did You Know?

Death that results from cardiac arrest is referred to as "Sudden Cardiac Death." It is estimated that there are as many as three-hundred thousand occurring each year, with the probability that most of these situations could have been reversed if prompt medical intervention could have been obtained.

Ectopy

The phrase "Ectopy" is taken from the term ectopia and could be defined as an abnormal position or behavioral occurrence of a body part or organ. When used in reference to the heart, ectopy would be described as abnormal myocardial behavior that manifests itself in various forms. The most common seen are referred to as premature ventricular contractions and premature atrial contractions. Although in most cases these anomalies are harmless, there is a danger of side effects if it occurs in the presence of heart disease or conditions such as Long QT Syndrome where there is a prolonged ventricular repolarization phase of the cardiac cycle. Fatal arrhythmias can be induced if premature ectopy enters into this vulnerable stage. Ectopic activity may be related to congenital defects, or it can be secondary to factors such as abnormalities present within the conduction system, medications, stress or fatigue, metabolic imbalance, excessive use of caffeine, nicotine, or alcohol. Ectopy may also be the result of conditions such as cardiac ischemia, which over time precipitates damage through oxygen deprivation This abnormal behavior can present in many different forms, a couple of the more common ones observed would include:

* Premature Ventricular Contractions or "PVC's" are caused by electrical impulses that originate from within the ventricles that induce a ventricular response before the impulse delivered by the sinoatrial node can follow its prescribed pathway. Premature ventricular contractions are generally seen as being bizarre in appearance with a large T-wave that is normally of the opposite deflection as that of the normal beat, that may be uni-focal and all having the same appearance, or multi-focal and possessing different morphologies. They may happen in single numbers, groups of two which are referred to as couplets, or three in a run which are referred to as triplets. Group beating that consists of numbers greater than three would classify them as ventricular tachycardia. They can happen every other beat which is referred to as bi-geminy, every third beat which is called tri-geminy, or every fourth beat which is referred to as quadra-geminy. Their occurrance can either cause the next normal cardiac cycle to be out of rhythm, or they can be accompanied by a "compensatory pause" that makes the allowance for and causes the next normal cardiac cycle to occur in the right timing.

* Premature Atrial Contractions or "PAC" Premature atrial contractions occur as the result of random electrical impulses that are initiated from origins within the atria that are outside of the prescribed conduction pathway. These impulses override the impulses delivered by the sinoatrial node inducing the atria to contract sooner than the normal time, causing them to be out or rhythm. As with premature ventricular contractions, premature atrial contractions can occur in configurations such as bi-geminy, tri-geminy, or quadra-geminy; but when they occur in group beating are referred to as either Supraventricular Tachycardia (SVT) which is a sustained episode of atrial group beating, or Paroxysmal Supraventricular Tachycardia (PSVT) that occurs in intermittent episodes.

* Premature Junctional Contraction or "PJC" This form of ectopy occurs when there is an irritable focus within the AV Junction that produces a premature beat. The beat will be retrograde as presented by the inverted P-wave that will commonly measure under 0.12 second that can be located before the QRS complex, within the QRS complex, or after the QRS complex. It will also produce a narrow QRS complex.

Edema

Edema is the swelling of bodily organs or tissue due to the retention of fluids within the interstitial spaces located outside the blood vessels. Under normal circumstances there is a balance between the fluid excreted from the capillaries that nourish the body's cells and the amount of the fluid that is reabsorbed back into the capillaries and drained by the lymphatic system. Edema occurs when there is an abnormality that precipitates an imbalance in the trade off between the excreted and reabsorbed fluid. One factor would be the presence of poor circulation caused by conditions such as congestive heart failure. Congestive Heart Failure, is a condition that is precipitated by diseases such as Coronary Artery Disease that damage the heart muscle and impair its ability to provide adequate blood movement to the body tissue and organs that are furthest from the heart. This causes a fluid build up within the blood vessels which increases the circulating volume and limits the amount of fluid reabsorbed, resulting in a fluid build-up in the subcutaneous interstitial spaces and tissue outside of the capillaries under the skin surface. Other medical issues that can contribute to this may include renal disease which compromises the kidneys effectiveness in removing an adequate amount of water and salt from the blood. Lymphadema, which is a problem with the drainage channels within the lymphatic system that affects its ability to adequately drain excess fluids. Pulmonary edema is a build-up of excess fluids within the lungs that is in most cases precipitated by stenosis of the hearts mitral valve or left ventricular chamber failure. Other types of edema would include:

* anasarca – this is a term that is used to describe a general swelling throughout the body that is caused by an increase in serous fluid within the tissue and cavities.
* dependent edema – also referred to as peripheral edema, this is an increase in the volume of extra cellular fluid in the limbs such as the legs and ankles that can cause swelling. This form of edema can be experienced at the end of the day or after being stationary or staying in a sitting position for long periods of time.
* mechanical edema – is a form of edema that is precipitated by outside sources such as restrictive or tight fitting clothing such as socks or undergarments that compromise fluid movement and removal.
* pitting edema - describes a pit or depression that is retained in the swollen tissue after pressing down on it. This type of edema is categorized into various levels of severity.

Did You Know?

A thin sac that is called the pericardium surrounds the heart. This protective envelope is made of a tough tissue that acts as a buffer to protect the heart from damage caused by contact between itself, and the lungs and chest wall. There is a fluid that is discharged between its smooth interior wall lining and the heart to act as a lubricant to protect it from friction caused by rubbing.

Electrocardiogram

The Electrocardiogram, which is also referred to as a 12 Lead, ECG, or EKG, is a non-invasive test that is used to record the electrical impulse as it travels through the heart. Although the first form of this test was performed in the year 1872 by Alexander Muirhead in St Bartholowmew's Hospital, the EKG was actually discovered in the year 1903 by a Dutch electrophysiologist named William Einthoven. Doctor Einthoven is also noted for labeling the various waves patterns observed on the EKG as the PQRST, and cataloging the characteristics associated with various cardiovascular abnormalities as observed on a wave pattern tracing. The use of the electrocardiogram as a diagnostic tool began within the United States starting in the year 1909. To perform this test a special gel is applied to the surface of the skin and 12 electrodes or metal plates are attached in various positions on the limbs and torso of the individual. These electrodes sense the electrical impulse as it travels through the heart and transmit the information to a receiver that is located within the computer where the signal is amplified by as much as 3000 times. The signals are then recorded by a heated stylus that causes a reaction on moving thermal graph paper producing 12 waveform tracings that present different views of the hearts electrical activity. Each tracing is then evaluated for proper cardiac wave form and interval time durations against prescribed guidelines that are known to be present in a normal healthy heart. With the help of this test the physician is able to recognize abnormalities that might be present within the heart such as:

* arrhythmia information related to abnormalities such as heart block. This information can indicate locations and severity levels.
* series EKG's are taken at the time of admit and are followed up at a later time to establish the location and extent of myocardial injury caused by a heart attack.
* to record and evaluate cardiac changes that might occur during information gathering procedures such as the exercise stress test. These tests are performed to evaluate the cardiovascular system while under stress for conditions such as ischemia and symptoms created by conditions such as Coronary Artery Disease.
* to detect imbalances in electrolyte levels such as potassium, magnesium, and calcium by observing abnormalities of the various wave forms.
* to evaluate pacemaker function.
* to recognize possible adverse physiological effects that may be induced by new cardiac drugs that are introduced into the body.

Did You Know?

The electrocardiograph (EKG machine) uses no ink to print tracings. There is a heated stylus that leaves an etching on special pre-printed thermal graph paper. This paper is coated with a substance that produces an imprint from stylus as it reacts to the heart's electrical activity.

"Tissue Plasminogen Activator" or "TPA" is composed of a protein similar to one that occurs naturally within the body. This substance is most effective within the first three hours of the onset of a heart attack or stroke for dissolving the clots responsible for the blockage of blood flow, but it does present an increased risk of bleeding. This chemical is also referred to as a "thrombolytic" or "clot buster" drug.

Embolism

An embolism is an blockage in one of the body's blood vessels caused by an embolus that was too large to pass through it. This obstruction can present itself as one of many different forms such as a blood clot, globule of fat, particle of tissue, or an air bubble. Embolisms usually occur in the medium to larger arteries like the pulmonary, cerebral and carotid arteries. They can also occur in areas associated with the intestines, arms, legs and kidneys. When a blockage occurs, the part of the body that is fed by the compromised artery is starved for oxygen rich blood. This precipitates a condition that is referred to as ischemia. If the blood is absent for too long a time period the organs and tissue affected by the blockage will suffer cellular damage or die through a process called necrosis. There are various types of embolisms, some of these would include:

* thrombo-embolism - this is a thrombus or blood clot that becomes lodged in a major artery stopping the flow of blood to the heart, lungs, or brain. The thrombo-embolism is the most common and is seen as the cause in most heart attacks and strokes.
* cerebral embolism - also referred to as an intra-cranial embolism. This event occurs when the blood flow to the brain is blocked as the result of a thrombo-embolism becoming lodged in a cerebral artery. This form of embolism is responsible for inducing strokes.
* arterial embolism - this occurs when an embolus becomes lodged in an artery, commonly at a fork. This event causes blood starvation and ischemic damage to the body parts beyond the block. The arterial embolism is seen more in situations involving arrhythmias such atrial fibrillation or where there is poor circulation due to the presence of existing heart disease.
* venous embolism - is caused by an air bubble that may have entered the body through a central intravenous line, a piece of bone marrow, or a globule of fat. This form of embolism is less common than the arterial embolism.
* pulmonary embolism - occurs when a blood clot originating from deep veins in the legs or pelvic veins travels through the circulatory system and blocks pulmonary arteries, affecting the flow of blood to the lungs.

Certain conditions put an individual more at risk of acquiring an embolism, a few of these would include:

* heart arrhythmias such as atrial fibrillation and atrial flutter where there is poor circulation due to abnormal heart function
* injury to the legs (deep vein thrombosis or "DVT")
* inability to move about due to illness or injury causing immobilization of limbs
* medications such as those used for hormone replacement therapy or birth control

Treatments for embolisms may include surgery, anticoagulation therapy through the utilization of medications such as heparin, coumadin, or the use of Tissue Plasminogen Activator (TPA) to dissolve clots. In the case of chronic recurring deep vein thrombosis a device known as a greenfield filter is surgically implanted to catch emboli before they can travel through the circulatory system.

First Degree AV Block

First Degree AV Block is a communication failure in the conduction system at the site of the AV junction or the His/Purkinje fiber system that precipitates a delay in the transmission of the electrical impulse as it is being transmitted from the atria to the ventricles. In this abnormality the impulse takes an extended period of time to execute a transmission within the guidelines of 0.12-0.20 one hundredths of a second from the beginning of the atrial depolarization to the point in the cardiac cycle when the ventricular depolarization begins. This abnormality is considered to be more of a delay as opposed to an actual heart block due to the fact that all of the electrical impulses do eventually make it to the ventricles. There are different classifications categorized by the level of severity according to the extended time duration for measurements that exceed these preset guidelines. The first level is considered slight and will measure within 0.21-0.24 one hundredths of a second. The second is classified as moderate with a time duration of 0.25-0.29 one hundredths of a second, with the third level referred to as severe possessing a measurement greater than 0.30 one hundredths of a second. Providing there are no other abnormalities present other than the prolonged time duration of the impulse transmission at the site of the AV node, the rhythm will have one atrial depolarization (P-wave) per ventricular depolarization (QRS complex), followed by the T-wave confirming ventricular repolarization. This abnormality may be caused by different factors such as diseased tissue within the AV node, myocardial infarction, electrolyte imbalance, or medications such as digitalis, beta-blockers, or calcium channel-blockers. First degree AV block is diagnosed with the utilization of an electrocardiogram (EKG) through measurements performed on a tracing that is printed on pre-graphed thermal paper. This device receives electrical signals that move through the heart at the body's surface through electrodes that are attached to the skin a different locations. These signals are then transmitted to a receiving unit that prints a tracing (EKG) capable of being measured on time incremented paper. This form of AV block generally produces no symptoms or heart related problems and can in most cases be treated with medication changes. It is estimated that First Degree AV Block is present in roughly about one person out of every thousand.

Digoxin (Digitalis) Antidysrhythmic, Cardiac Glycosoide

This drug is comprised of a chemical that is extracted from a plant called the "Digitalis Lanata." Digoxin enhances the power of the hearts contractions, providing for greater pushing force and increased blood perfusion to the body. Its properties also regulate the heart rhythm by decreasing the speed of the conduction at the site of the atrioventricular node. These properties were discovered in the year 1785 when William Withering, who was a British scientist, found it useful in treating certain diseases of the heart. Digoxin needs to be monitored for possible side effects related to dysrhythmias such as AV heart block or drug induced bradycardia.

Heart Attack

The heart attack, also called a "Myocardial Infarction," is an acute coronary event that occurs when the heart muscle becomes starved for oxygen-rich blood. This is commonly the result of reduced blood flow that may be induced by different factors. It can be acute in nature and caused by a sudden blockage from a blood clot that has become lodged within a coronary artery, or it may have been an event that was precipitated by a degenerative disease that over time reduced blood flow such as Coronary Artery Disease. Coronary Artery Disease, which is also referred to as "CAD" or "artherosclerosis," is a build-up of plaque within the arteries that narrows the passageways (lumens) and reduces the amount of blood flow to the body and its organs; in this instance the heart. This diminished supply of blood causes the heart to experience necrosis (cellular death) through ischemia which is a defined as the inadequate supply of oxygen-rich blood that is being delivered to the body's tissue or organs to satisfy a needed demand. The myocardial cells become affected by ischemia within the first 10 seconds of diminished blood flow caused by the blockage and will suffer necrosis in a matter of minutes to the part of the heart muscle that is being fed beyond the compromised artery suffering the damage if not revascularized. It is estimated by the American Heart Association that the average age of an individual suffering their first heart attack is in their mid 60s. It is also estimated that in the year 2006 approximately 700,000 individuals will have suffered a heart attack, with another 500,000 experiencing recurrent heart attacks.

Heart attacks in most cases will produce symptoms, however, they can occur without displaying symptoms. These episodes are referred to as "silent heart attacks" and are normally discovered during an electrocardiogram (EKG) at a later date. Silent heart attacks are sometimes seen in individuals such as farmers who attribute the angina to muscle soreness resulting from a hard days work. When there are symptoms present, they would commonly include chest pain, pressure, or squeezing that may radiate to the jaw or arm. This is referred to as angina. These symptoms may also be accompanied by shortness of breath, episodes of pre-syncope and syncope, nausea, fatigue and sweating.

There are many factors that can contribute to the onset of a heart attack, some that may include:

* uncontrolled high blood pressure causes left ventricle hypertrophy, a condition known as remodeling. This increases the chance for clot formation due to the lack of blood movement caused by poor circulation.
* elevated LDL (bad cholesterol) and glucose levels
* infections such as influenza have been shown to increase the chances of a rupture in the plaque that has accumulated within the arterial lumens.

The heart attack is diagnosed in a couple of different ways. The initial methods utilized by hospital personnel after the individual presents with symptoms that are indicators of a possible myocardial infarction such as angina, elevated blood pressure, increased heart rate, jugular vein distention, shortness of breath and nausea, electrocardiogram (EKG) and blood lab work. The electrocardiogram will display different findings to the Physician such as elevation or depression in the ST segment of the heart rhythm tracing with relationship to the isoelectric baseline, pericarditis, hypertrophy and any arrhythmia activity present within the conduction system. It will also provide the specific location of the heart muscle that is suffering the attack and to what degree it is being affected. Blood work would primarily involve the CPK, CKMB, and troponin levels and is useful in diagnosing both Q-wave (with ST segment elevation) and non Q-wave (without ST segment elevation) myocardial infarctions. These blood tests are referred to as "cardiac enzymes" and are specifically related to cardiac cellular damage. Substances called creatine kinase, isoforms and troponin (which are also called cardiac markers), are released into the bloodstream when cardiac cell membrane becomes ruptured during a heart attack and myocardial cells die. The serial cardiac enzyme test measures the amount of these substances within the blood over a period of 24 hours. They are referred to as the CPK, the CK MB test which starts to become elevated about 6 hours after the onset of a heart attack and the Troponin. The troponin test will also show protein levels that are not detected by the CK-MB test that can indicate a myocardial infarction of smaller magnitude. Elevated levels of these chemicals present in this blood test are an indication that a heart attack has occurred and depending on how sharp and high the levels have increased will help in determining what magnitude it was. It is necessary however, to evaluate all three of these enzymes together to achieve a true result.

Normal Cardiac Enzyme Levels:

CK 1 Men 25-130 micrograms per liter
Women 10-150 micrograms per liter

CK 11 up to 7 micrograms per liter
This will display a rise in 3-6 hours, peak in 12-24 hours and stay in the blood for a total duration of 1-3 days

Troponin up to 0.4 nanograms per milliliter
This will display a rise in 3-6 hours, peak in 12-16 hours and be present within the blood for a duration of 2 weeks.

* The cardiac troponin can rise to 40 times its normal level.
* The CK MB can rise up to 5 times its normal
* The CK MB is used in evaluating whether the chest pain is being caused due to a heart attack or from another source.

Did You Know?

The first published recording of an acute myocardial infarction occurred in the year 1920 by a Dr. Harold Pardee in New York. The tracing displayed a tall T-wave that began well up the R-wave.

Heart Block

Heart block is an abnormality or defect that is present within the heart's conduction system that precipitates a delay or complete failure in the transmission of the electrical impulse as it travels along its prescribed pathway from the sinoatrial node to the purkinje fibers. The block can occur in the sinoatrial node as it generates the impulse or as the impulse exits the sinoatrial node en-route to the atrium. It can be a delay or complete communication failure as the electrical impulse is transmitted from the atria to the ventricles through the site of the atrioventricular node. Or it can be a delay in either the right or left bundle branch located in the ventricles beyond the AV Junction. In the normal conduction cycle an impulse originates in the sinoatrial node and travels along a prescribed conduction pathway to initiate systematic and predictable muscular contractions in a timed order. In heart block the impulse is blocked or delayed, altering its ability to conduct myocardial tissue along its normal conduction pathway. This can result in an abnormally slow heart rate or in the case involving complete block where there is no impulse to the ventricles, cardiac standstill can occur. Heart block can be caused by various factors such as:

* damage to the myocardium through cardiac ischemia, which is a lack of oxygen-rich blood to the heart.
* hyperkalemia which is when there is an elevated level of potassium within the blood.
* hypokalemia which is a low potassium level within the blood.
* degeneration of the hearts conduction system that sometimes occurs with the aging process.
* diseased tissue in the sinoatrial or atrioventricular node.
* post heart surgery.
* certain cardiac medications such as beta-blockers.
* a congenital condition that has been present since birth.

There are different classifications of heart block that are directly associated with their origin and characteristics, they include:

Sinoatrial Block, which is also referred to as "SA Block." This is a block that is precipitated by a disturbance in the conduction system that originates at the site of the sinoatrial node. Complications associated with cardiac drugs or disease disrupt the formation of the electrical impulse or impede it as it exits the sinoatrial node en-route to the atria. This action precipitates a failure of the electrical impulse to initiate one or more complete cardiac cycles. This arrhythmia is commonly transient in nature, entering and exiting a rhythm in short episodes that will be displayed on the tracing of an electrocardiogram (EKG) as the absence of one or more complete cardiac cycles (PQRST). In retrospect as compared to the atrioventricular block (AV block) where there is an atrial depolarization without a ventricular depolarization (P-wave without QRS), this block born at the sinoatrial level of the myocardium fails to produce an entire cardiac cycle (no P-wave or QRS). For this reason, this form of heart block is also referred to as a sinus pause. Possible causes that may contribute to the onset of this arrhythmia can include diseases such as sick sinus syndrome, an inferior wall myocardial infarction, or toxicity from drugs such as potassium, quinidine, acetylcholine, beta blockers, or glycosides.

"AV Heart Block" is a communication failure in the conduction system that precipitates a delay in or complete failure of the electrical impulse as it is being conducted from the atria to the ventricles. There are various types atrio-ventricular heart block, they are:

1st Degree AV Block - This form of heart block causes a delay in the impulse as it is transmitted from the atria to the ventricles causing it to exceed the prescribed 0.20 second time duration as seen in normal sinus rhythm. Although this is classified as a heart block, it is considered more of a conduction delay due to the fact that every impulse sent eventually conducts the ventricles. This AV block generally produces no symptoms or health related problems and in many cases can be treated with medication changes. It is estimated that First Degree AV Block is present in roughly about one person out of every thousand. This form of heart block has different classifications with reference to the extent of the time duration involved. These are described as follows:

* 0.21 – 0.24 is slight
* 0.25 – 0.29 is moderate
* 0.30 and greater is considered severe

2nd Degree AV Block - This form of heart block which is also referred to as "Type 1 and Type 2" is precipitated by a failure in the communication between the atria and the ventricles that is precipitated at the site of the atrioventricular node and bundle of his. This intermittent blockage or delay in the transmission of the electrical impulse as it travels through the atrioventricular node to the ventricles can be caused by a variety of different reasons, some of these reasons may include:

* diseased tissue in the atrioventricular node or bundle of his
* cardiac medications such as calcium-channel blockers and beta-blockers
* the result of a prior heart attack
* an imbalance in the levels of the electrolytes
* inflammation of the myocardium – swelling in part of the heart muscle caused by injury, irritation, or infection
* cardiac ischemia – the lack of oxygen-rich blood supplied to the heart muscle commonly caused by narrowing arterial pathways due to plaque build-up
* natural aging of the hearts conduction system
* condition present from birth
* a response or change in the myocardium resulting from a prior heart surgery

There are two different forms of second degree atrioventricular heart block which possess different characteristics, they are classified as "Type l" and "Type ll". The following is a brief description of each:

Second Degree AV Heart Block "Type l" – this variant of AV block is also referred to as "Mobitz l" or "Wenckebach." The electrical impulses that are communicated to the ventricles from the atria in Type 1 are progressively delayed with each cardiac cycle, producing a lengthening PR interval with every beat. When the impulse reaches the relative refractory period in the cycle, the ventricular depolarization will drop, displaying a P-wave without a QRS on the EKG tracing. This block is transient in nature inserting itself in and out of another rhythm in brief episodes.

* Diagnosis – this arrhythmia is diagnosed through the taking of an electrocardiogram (EKG) to record the hearts electrical activity.
* Treatment – medication changes or the utilization of a pacemaker.

Second Degree AV Heart Block "Type ll"- this arrhythmia causes an intermittent block at the AV junction that prevents select electrical impulses from being transmitted to the ventricles. This arrhythmia possesses a conduction block that can be regular such as 2:1, or can be variable with changing conduction ratios. Unlike 2nd Degree "Type 1" which forewarns of the missed beat with a gradual increase in the PR interval, the onset of 2nd Degree "Type ll" is sudden and without warning. The PR intervals will all be uniform and the P-P's and R-R's will march out perspectively, but due to the potential for the occurrence of variable conductions this block can have varying R-R measurements.

* diagnosis – this arrhythmia is diagnosed through the taking of an EKG (electrocardiogram) to record the hearts electrical activity.
* treatment – this form of second degree heart block is of a more serious nature than second degree type l, with the possibility to a progression to a complete heart block. In severe cases medications such as atropine can be used to increase the heart rate, or a permanent pacemaker can be implanted to manage the hearts conduction system.

Third Degree AV Heart Block – 3rd Degree AV Block is also referred to as "Complete Heart Block" and is a form of "AV Dissociation." In this block there is the complete lack of communication between the atria and the ventricle at the site of the AV junction. As with 2nd Degree Type ll, this arrhythmia's P-P's and the QRS complexes will march out, but unlike the 2nd Degree Type ll, will be completely dissociated from each another. Due to the fact that the electrical impulses from the atria never make it through the AV junction, the ventricular heart rate will be at its intrinsic rate of 20-40 beats per minute. This low heart rate can produce symptoms caused by the diminished cardiac output such as fatigue, confusion, syncope and pre syncope, chest pain and shortness of breath. 3rd Degree AV Block can be induced by various conditions, some may include:

* cardiac ischemia
* hyperkalemia – elevated potassium level
* hypokalemia – low potassium level
* heart surgery
* aging process of the conduction system
* congenital defects
* cardiac medications such as beta-blockers

In many situations this arrhythmia is treated with a permanent pacemaker to manage the conduction cycles and regain uniformity to the atria and ventricles.

Bundle Branch Block (BBB) - The bundle branch block is a delay or complete blockage in the hearts conduction system that occurs below the atrioventricular node. This defect impedes the electrical impulse sent from the sinoatrial node as it travels along this pathway through the myocardium to the purkinje fibers in the ventricles to complete the conduction cycle. In this block the abnormality is precipitated by compromised tissue located at the site of the right or left Bundle Branches where they separate following the Bundle of His, with the extent of the block ranging from a slight conduction delay to complete failure of the impulse to pass through. Diagnosis of this condition is commonly done with the use of a test called an electrocardiogram or EKG, which records the electrical activity as it moves through the heart. This test produces a tracing that is used to measure the QRS to ensure that the measurements are within the set guidelines of 0.06 – 0.10 sec. as seen in normal sinus rhythm.

Measurements greater than this set limit indicate that there is a prolonged time duration experienced at this point in the conduction cycle, confirming the presence of a bundle branch block. Once a Bundle Branch Block has been established, there are a variety of tests to help a physician to determine the actual cause and to decide on the proper treatment. This abnormality is sometimes present for unknown reasons, but can also be provoked by issues related to other medical conditions or problems. Some of these may include:

* chest injury
* myocardial damage from a prior heart attack
* deterioration of the conduction system from aging
* hypertension – high blood pressure
* congestive heart failure
* chronic obstructive pulmonary disease (COPD)
* an abnormality present from birth / congenital

This defect will generally produce no symptoms, however, episodes of dizziness or fainting (syncope) may occur from the diminished oxygen-rich blood supply caused by too slow of a heart rate. In severe situations a pacemaker is surgically implanted into the individual's chest to manage the hearts conduction cycles and maintain a set limit that the heart rate will not drop below As with other abnormalities associated with the heart, the symptoms produced by this defect will be amplified in the presence of cardiac disease.

<u>Inherent Intrinsic heart rates:</u>

Sinoatrial node	60-100 beats per minute
AV Junction	40-60 beats per minute
Purkinje Fibers	20-40 beats per minute

Did You Know?

The electrocardiograph (EKG) was first invented in the year 1903 by an electrophysiologist named William Einthoven.

Death that results from cardiac arrest is referred to as "Sudden Cardiac Death". It is estimated that there are as many as three-hundred thousand occurring each year, with the probability that most of these situations could have been reversed if prompt medical intervention could have been obtained.

Holter Monitor

The Holter Monitor which is also called an "Ambulatory EKG," was invented in the year 1949 by an American biophysicist named Dr. Norman Holter. This monitor is designed to be worn for a period of 24 hours or more by an individual that is experiencing chest pain, palpitations, or other symptoms that may be cardiac related. Its purpose is to monitor and record the hearts electrical activity such as average heart rate, high and low heart rates and ectopy such as premature ventricular contractions or premature atrial contractions while the individual performs their normal daily routines. The holter monitor, depending on the model worn, utilizes from 3 to 8 electrodes that are attached to the torso of the individual. The electric signals that travel through the heart are picked-up by these electrodes and are transmitted to the monitoring device through leads (wires) that records and maintains a historical record to be evaluated at a later time. The record of any arrhythmias that might occur during the wearing of this device can be matched with notes kept by the individual of the times of the activities such as sleeping or exercising being performed by the wearer during the event. This device provides valuable information such as the affects of the implementation of new cardiac drugs and to monitor the heart's activity after a heart attack. It is also incorporated to observe the function of a new pacemaker after its introduction to manage the individual's conduction system.

Did You Know?

Epinephrine

This is a synthetically produced drug that mimics the hormone released by the medulla of the adrenal gland. This chemical stimulates the heart giving increased muscular strength and endurance. Its use is indicated in situations where the use of electrical shock to stimulate myocardial response will have no effect such as asystole when the heart is in a depolarized state; or where there are electrical impulses present but no mechanical myocardial function (PEA). Adrenalin (Epinephrine) is a chemical that is secreted by the medulla of the adrenal gland. This hormone was first isolated and named in the year 1901 by a U.S. chemist named Dr. Jokichi Takamine to be present in the secretions from animal adrenals. Adrenalin is normally released in times of fear or injury to prepare the body for physical or mental stress, producing what is referred to as the body's natural "Fight or Flight" response. Today, this chemical is synthetically manufactured to produce the same physiological changes. This drug produces an immediate increase in the resting energy level (basal metabolism) and blood sugar level, producing enhanced myocardial contraction strength and endurance, which increases the individuals heart rate, blood pressure, and perfusion to the body. For this fact it is administered to individuals during episodes of cardiac arrest, asystole and PEA (pulseless electrical activity) to stimulate a myocardial response.

Hypertension

Blood pressure readings are considered normal if at or below 120mmHg systolic and 80mmHg diastolic, with an optimum reading at 115mmHg for systolic and 75mmHg for diastolic. Hypertension is described as a consistently high blood pressure that presents with readings at or above 140mmHg for systolic and 90mmHg for diastolic. The systolic pressure reading is taken to evaluate the arterial pressure as the heart contracts and forces its fill of blood into the circulatory system, with the diastolic reading representing the pressure while the heart is in between contractions and the blood is being returned to the heart through the circulatory system. Hypertension can be referred to as primary which is also called essential, or it can be secondary and meaning that it is the result of another condition. Essential hypertension is usually found to be present with no medical reason that would have produced compromising factors to induce or precipitate the condition. The term secondary hypertension is used when there is the presence of a medical condition or disease that has been found to have influenced the elevation in the blood pressure. These potential secondary reasons are checked for in tests such as the electrocardiogram which will show EKG changes or a chest X-Ray to show an enlarged heart muscle that can indicate an overworked heart. The Physician will also evaluate results from blood tests such as electrolyte, glucose, creatnine and cholesterol for elevated levels. These results will provide the Physician with information regarding the kidney function, the possible presence of diseases such as Coronary Artery Disease and the levels of minerals such as potassium, sodium and calcium within the blood. Due to the fact that hypertension does not display symptoms until the pressure becomes extreme, it is commonly diagnosed through the observation of elevated readings during routine medical examinations. It is therefore given the title "The Silent Killer." Blood pressure readings that reach 240mmHg or higher for systolic and 120mmHg or higher for diastolic are classified as accelerated hypertension and can start to display symptoms such as headaches, fatigue, blurred vision, or nausea. When there is the presence of accelerated hypertension that has reached the point where there is organ damage that is occurring or that damage is imminent, this condition is classified as a hypertensive emergency. When this situation is present in addition to the presence of high blood pressure induced intracranial pressure it is called malignant hypertension. Accurate blood pressure readings depend when and how the pressure is taken. Caffeine produces an artificial and temporary elevation in blood pressure readings and the test should not be performed for at least 1 hour after its consumption. Other factors that may influence the readings would include certain drugs used in cold medications, the size of the blood pressure cuff with relation to the size of the arm and the individuals present stress level. The effects created by hypertension on the body can produce damage that may be acute and temporary or chronic to the organs such as the heart, kidneys, eyes and brain. The increased pressure within the arteries can cause aneurysms or weak locations in artery walls to rupture, induce heart attacks and strokes, cause retina deterioration and precipitate irreparable renal damage. High blood pressure in a chronic state can cause heart failure by creating an extra workload on the heart muscle, precipitating a condition that is referred to as cardiomyopathy.

Idioventricular Rhythm

The Idioventricular rhythm is considered to be a ventricular escape rhythm that functions through myocardial cellular automaticity that acts as a back up to prevent complete cardiac standstill. This escape mechanism occurs in situations when the sinoatrial node fails to generate an electrical impulse above the ventricles. There are two different classifications of this rhythm that are referred to as idioventricular and accelerated idioventricular; with the difference between the two being the heart rate. The normal idioventricular rhythm occurs when heart rate is between twenty to forty beats per minute, which is the intrinsic heart rate of the ventricles. Accelerated idioventricular, also known as "AIVR" is precipitated by enhanced ventricular automaticity that possesses a heart rate between forty-one to one-hundred beats per minute. This rhythm can be caused by various factors such as a heart attack which usually produces this rhythm within the initial twenty-four hours of its onset, or it may be due to various other reasons. A few of these reasons would include heart disease such as cardiac ischemia, drug toxicity from medications such as digoxin, metabolic imbalance, hypoxemia or a subarachnoid hemorrhage which is an event that occurs when there is bleeding around the brain caused by a ruptured aneurysm or head injury. It was also found after myocardial infarctions that idioventricular rhythm was present in greater than 15% of the Percutaneous Transluminal Coronary Angioplasty (PTCA) procedures during the re-profusion of the cardiovascular system. Treatment for this rhythm would concentrate on dealing with the symptoms created and in correcting the underlying reason that precipitated it. Medical intervention would not be necessary for patients without symptoms, however for patients who present with symptoms precipitated by low blood volume caused by decreased perfusion there are a couple of different treatment methods. One method would include the use of medications such as atropine which would be introduced into the system to stimulate and enhance the sinoatrial node and myocardial activity. Another option to increase the blood volume perfused to the organs would be to control the heart rate with the utilization of transcutaneous pacing to manage the hearts conduction using electrical impulses generated by an external pacemaker. This form of pacing is used at the bedside in emergency situations where patients present with symptoms and is performed by placing electrodes on specific locations on the torso that receive and transmit timed electrical impulses from an Automated External Defibrilator (AED) device that are standard equipment on a hospital code cart or other external means.

Did You Know?

Being overweight increases the risk of becoming afflicted with heart disease. It is estimated that sixty percent of the adults in the United States are obese and that forty percent engage in minimal or no physical activity. Studies have also shown that obesity in children has almost tripled in the past fifteen years.

Junctional Rhythm

The junctional rhythm is an escape mechanism that originates in the AV junction. In the junctional rhythm the electrical impulses that would normally be delivered from the sinus node would originate at the site of the AV junction. This site contains the atrioventricular node that is the electrical impulse transmission point that is located between the atria and the ventricles which normally acts as a regulator and transmission point for the electrical impulses that are sent by the sinus node to the atria, that are eventually conducted to the ventricles. This junction can produce its own electrical impulse in emergency situations to ensure myocardial depolarization through a process called automaticity. This will occur when there are episodes of extreme bradycardia such as those seen in heart blocks, a heart rate produced by the sinus node is less than rate of the AV node such as those seen in sick sinus syndrome, or when there is increased automaticity within the AV junctional tissue. Some other possible reasons or conditions that may precipitate this arrhythmia can include cardiac surgery, digitalis poisoning, myocardial infarction, or glucose toxicity. The junctional rhythm produces different heart rates that would be categorized into the following classifications:

* junctional bradycardia - 20 to 40 beats per minute
* junctional - 40 to 60 beats per minute
* accelerated junctional - 60 to 100 beats per minute
* junctional tachycardia - greater than 100 beats per minute

This rhythm is identifiable on an electrocardiogram (EKG) tracing through the observation of an inverted P-wave. This anomaly is caused by retrograde conduction that originates from the AV node and travels in a direction away from the positive pole. When there is an inverted P-wave before the QRS cycle and the measurement in normally less than 0.12 one hundredths of a second, it would indicate that the atria depolarized before the ventricles and that a retrograde conduction has occurred. If there is no visible P-wave at all (can be hidden in the QRS complex), this means that the atria and the ventricles depolarized simultaneously. When there is an inverted P-wave that follows the QRS cycle it indicates that the ventricles depolarized before the atria. However the rhythm must be regular along with these findings so as not to be confused with a fine atrial fibrillation when the junctional p-wave is not visible. Treatments in cases involving the junctional rhythm are not normally necessary due to the fact that it commonly produces no symptoms. However, treatment used in caring for patients that experience symptoms such as a bradycardic heart rates induced by this arrhythmia (junctional bradycardia) might necessitate the use of medications or the implementation of a pacemaker to manage the hearts conduction.

Did You Know?

The electrocardiograph (EKG machine) uses no ink to print tracings. There is a heated stylus that leaves an etching on special pre-printed (graph) thermal paper. This paper is coated with a substance that produces an imprint from the heat generated by the stylus.

Mitral Valve Prolapse

Mitral Valve Prolapse or "MVP," is a congenital abnormality that affects the hearts mitral valve. This is a lifelong condition that is believed to be the result of disturbances that have altered the development and formation of the valve during the gestation period. This defect is also referred to as "Click-Murmur Syndrome" due to the sounds heard through the stethoscope during the cardiac cycle. When opening and closing the defective valve makes a "click" sound, with an audible "murmur" sound made by the regurgitation that takes place during transfer of blood from the left ventricle to the body. The mitral valve has two triangular flaps and is located between the left atrium and left ventricle. Its function is to open and allow the blood to enter the ventricle from the atria, then to close to form a positive seal against the pressure created within the ventricular chamber during its depolarization. This defect becomes a problem when regurgitation due to a bad seal of the leaflets allows them to open upward into the atria during the ventricular depolarization. This allows the regurgitation of blood from the ventricles back through the valve into the atrium. In severe cases the regurgitation decreases the blood volume sent to the body by allowing less force to move the blood during the left ventricular perfusion. Due to the additional workload to the heart muscle to provide adequate circulating blood volume to replace that lost through regurgitation and perfusion, this situation can eventually lead to conditions such as congestive heart failure. Although it is not readily recognizable in most people and can be lived with, some individuals may experience palpitations, fatigue, lightheadedness, shortness of breath, chest tightness, or chest pain. These symptoms usually prompt them into having a more in-depth examination by their physician. Diagnosis of this condition is possible through the observation of the audible click-murmur sounds heard through the stethoscope during the systolic phase. It can also be diagnosed by using echo technology through the utilization of an echocardiogram which produces high frequency sound waves that are transmitted toward the area under observation with a devise called a transducer. These sound waves are then reverberated back to become an image on a monitor to be viewed by the Physician. Echocardiogram technology makes it possible to observe the heart as it functions, allowing for the Physician to see malfunctions as they occur. In situations where this defect causes severe regurgitation that produces symptoms, the valve is repaired or replaced through an invasive surgical procedure. In situations where there is minimal regurgitation but no symptoms present, antibiotics are administered to decrease the possibility of infection within the heart; a condition referred to as endocarditis. It is estimated that three to four percent of the population have mitral valve prolapse and that it is non-discriminatory in either women or men, affecting both sexes equally.

Did You Know?

If all of the resting phases of your heart that occur between beats were combined, there would be a twenty-year period that your heart was idle.

Normal Sinus Rhythm

In normal sinus rhythm the electrical impulses are formed in the sinoatrial node located in the upper right atrium and follow the prescribed conduction pathway through the heart muscle, initiating a firing rate between sixty and one-hundred beats per minute. The electrical impulse is generated within and exits the sinoatrial node. It then enters into an inter-atrial pathway that spreads the electrical impulse from the right and left atria, causing them to contract (depolarize) simultaneously. The electrical impulse then proceeds along an intra-nodal pathway where it enters the atrioventricular (AV) node. At this point the impulse is momentarily delayed to allow for ventricular filling time as the atria empties its fill of blood into the ventricles. It then proceeds through the atrioventricular node, through the bundle of his, down the right and left bundle branches to the purkinje fibers where it causes the ventricles to contract (depolarize). The rhythm will have measurements that are constant and within the prescribed limits. The PRI will be within 0.12-0.20 one hundredths of a second from the start of the atrial depolarization to the start of the ventricular depolarization., the QRS will measure within 0.06-0.10 one hundredths of a second from the start to the completion of the ventricular depolarization and a QT interval will measure within 0.36-0.44 one hundredths of a second to represent the time duration from the start of the ventricular depolarization to the finish of the ventricular repolarization. The rhythm will be regular in the R-R measurements and will have one atrial depolarization for each ventricular depolarization. There will be one upright and rounded P-wave for each QRS complex, with an upright T-wave which represents ventricular repolarization. This constitutes the established proper guidelines for the heart's intrinsic normal sinus rhythm that provides a basis to use as a reference when interpreting proper or improper myocardial behavior.

Did You Know?

A thin sac that is called the pericardium surrounds the heart. This protective envelope is made of a tough tissue that acts as a buffer to protect the heart from damage caused by contact between itself, and the lungs and chest wall. There is a fluid that is discharged between it's smooth interior wall lining and the heart to act as a lubricant to protect it from friction caused by rubbing.

After the implementation of a pacemaker, certain information may be obtained over the telephone through a process called "transtelephonic monitoring". By utilizing a special device, information that is gathered and stored within the pacemaker can be transmitted from the individuals residence to the physicians office. During this transtelephonic evaluation the battery life and the programming can be checked to ensure proper function.

Pacemaker

The pacemaker is a device that is designed to produce and deliver a programmed electrical impulse to induce a myocardial reflex action in situations where there is an abnormality or failure in the hearts intrinsic conduction system to provide stimulation to induce its own depolarization. It consists of a pulse generator containing a microprocessor and lithium battery, wire leads and electrodes that are used to transmit the electric impulses. This unit is hermetically sealed within a titanium housing due to its sealing properties, ability to shield its components from electromagnetic interference and limited active properties that might otherwise make it conflict with the body's immune system. To perform its function, the pacemaker pulse generator creates and electrical impulse that is transmitted through the lead wire to an electrode that is placed in the selected chamber within the myocardium to provide stimulation. At this point the impulse is conducted by the chambers muscular tissue and a reflex action occurs in the form of a contraction. This action forces the blood to be moved from the chamber to its next destination within the cardiac cycle. This device can be permanently implanted into the individual's chest which is referred to as a "permanent pacemaker," or employed temporarily at the bedside in emergency situations. This form of pacing is referred to as bedside pacing, temporary pacing, or external pacing. There are a variety of different types of pacemakers that are designed to be programmed to perform different functions, with the selection dependant upon the medical situation. A few examples of the various designs available would include:

* bi ventricular – this particular unit came about in the year 2002 to be used in the treatment of conduction abnormalities caused by irreversible medical conditions such as Congestive Heart Failure. It is designed to restore myocardial perfusion by sending an electrical impulse to both of the ventricles simultaneously to induce a synchronized depolarization so that of all ventricular chamber walls contract simultaneously.
* on demand – this style was developed in the mid 1960s and is programmed to maintain a heart rate above a prescribed number of beats per minute. This unit senses the intrinsic activity of the heart and it operates only when the heart rate falls below the pre-set limit.
* single chamber – this pacemaker came about in the 1960's and is designed to deliver an electrical impulse to either the atrial or the ventricle individually.
* dual chamber – first used in the 1970's, this pacing mode is designed to mimic the electrical impulse stimulation as it would be delivered by the sinoatrial node by sending both the atria and the ventricle a programmable timed impulse.
* rate modulated – this unit is also referred to as "rate responsive" and was developed in the 1980s. The rate modulated pacemaker is designed to adjust the heart rate according to the activity level of the individual by sensing bodily changes such as an increase in the respiration rate.
* fixed rate - these are programmed to send continuous electric impulses to achieve and maintain a particular heart rate without regard to any other factors.

* transcutaneous pacing – this is a form of temporary or external pacing in which pads are placed on the outer chest wall to receive and transmit electrical impulses from a pulse generator located outside of the body. This type of pacing is done in medical emergencies that might involve hemodynamically unstable situations such as extreme bradycardia.

Pacemaker Terminology

AV Delay the time delay between the electrical impulse sent to depolarize the atria and the electrical impulse sent to depolarize the ventricle.
VA Delay the time duration between the ventricular depolarization in the previous cycle and the atrial depolarization in the new cycle.
Pulse Interval the total of the AV and the VA intervals in dual chamber pacermakers as measured in milliseconds.
Capture confirmation that an electrical impulse sent to the atria or ventricles has stimulated a depolarization within the proper location and timing.
Crosstalking this occurs when the impulse sent to the ventricles from a dual chamber pacemaker is retrograde conducted to the atria inducing a tachacardic heart rate.
Dual Chamber Pacing the sending of an electrical impulse to both the atria and the ventricles.
Electromagnetic Interference (EMI) interference that is caused by outside electrical or magnetic equipment that can interrupt the normal operation of the pacemaker.
Electrical Output the amount of electrical current required to initiate a depolarization.
Fusion Beat this happens when the impulse sent by a pacemaker and the hearts own (intrinsic) beat happen simultaneously.
Hysteresis a prolonged pulse interval to allow the heart to provide it's own stimulation.
ICD (Implantable Cardioverter Defibrillator) this is a pacing device that is implanted within the chest that is designed to control rapid heart rates and to terminate ventricular fibrillation, ventricular tachycardia and torsades de pointes by delivering an electrical shock to reset the hearts natural pacemaker (SA node).
Pulse Generator the part of a pacemaker that is designed to generate the electrical impulse that is used to deliver a shock to the heart muscle.
Rate Modulated pacemaker that senses the activity level of an individual and makes the heart rate adjustments automatically.
Single Chamber Pacing the sending of an electrical impulse to either the atria or the ventricle.
Standby Rate the lowest heart rate that a pacemaker will allow the intrinsic heart rate to fall to before sending impulses to initiate pacing.

Did You Know?

The first portable battery powered pacemaker was developed on the year 1957. This unit functioned with transistors which eliminated the fear of losing public utility electricity needed to power it.

Pacemakers that were equipped with leads that discharged steroids became available in the 1980s. These new leads reduced myocardial inflammation at the site of the lead placement.

Pericarditis

The pericardium is a thin sac of fluid that surrounds the heart and acts as a buffer to protect it from contact between itself, the lungs and the chest wall. Pericarditis is a condition that occurs when outside factors cause the sac to become inflamed or to become enlarged due to excessive fluid build-up. Pericarditis can sometimes present with no known reason, however there are known causes that precipitate this condition. Some of these would include the strep virus, influenza, HIV, kidney failure, cancer, hepatitis B and damage from prior heart attack. This event can also occur due to a trauma such as an accident, puncture or stab wound. This opening in the pericardial sac would allow fluids from the outside to enter causing an excess fluid build-up. The constriction precipitated by this condition creates pressure against the surface of the heart known as Cardiac Tamponade, which is estimated to be one of the results of Pericarditis that presents in fifteen percent of reported cases. Pericarditis can present itself in either one of two forms. It can be chronic and consisting of episodes with long duration and frequent recurrences, or it may be acute in nature with sharp and severe episodes that happen suddenly with rapid fluid build-up. Acute Pericarditis is the more commonly seen of the two. Complications from this condition not only compromise the flow of oxygen-rich blood to the coronary arteries located on the surface of the heart, but also restrict the expansion of the heart during the refilling phase of the cardiac cycle. This limits the fill of blood, resulting is a decrease in the volume of blood perfused to the body and lungs which can cause various symptoms and medical problems. Some of these may include progressively worsening chest pain, rapid heart rate, shortness of breath, weakness, fatigue and ischemia. Another possible result that could become an issue would be the additional workload put on the heart muscle as it attempts to compensate for the decreased oxygen-rich blood supply. This may eventually lead to a condition called heart failure. Diagnosis would include the evaluation of the presenting symptoms such as chest pain that increases and decreases with positional changes, becoming more intense when the individual lays down and less intense as they sit and lean forward. Electrocardiogram (EKG) tracings would show depression of the PR segment and elevation of the ST segment. These findings would be accompanied by a heart rub or "Friction Rub" sound that is audible through a stethoscope.

Treatment for this condition is dependant upon the diagnostic findings. Some of the possible alternatives that may be employed can include:

* a procedure referred to as Pericardiocentesis which would be used in cases where there is the presence of Cardiac Tamponade. This is done to introduce devices into the pericardium to promote fluid drainage to relieve pressure.
* the use of anti-inflammatory medications such as steroids and antibiotics to reduce and keep inflammation from recurring.

Did You Know?

The phrase "Ectopy" is taken from the term ectopia, which is defined as an "abnormal position of a body part or organ." When used in reference to the heart, ectopy could be described as abnormal myocardial behavior manifesting itself in various forms, the most commonly seen referred to as premature ventricular contractions and premature atrial contractions.

Peripheral Artery Disease

Peripheral Artery Disease, which is also referred to as "PAD" or "Leg Artery Disease," is a condition that is precipitated by diminished blood flow to the legs. Over the years fatty deposits and plaque build-up (atherosclerosis) within the arteries causing a reduction in the size of the lumens within the blood vessels. This decreases the volume of blood that is supplied to the legs creating a condition called ischemia, which is oxygen deprivation to bodily tissue that over time can result in necrosis, or tissue death. Peripheral Artery Disease is precipitated mainly by atherosclerosis, but there are outside factors that can increase the possibility of becoming afflicted such as obesity, smoking, high blood pressure, or diseases such as diabetes. In its early stages Peripheral Artery Disease does not always display symptoms, but due to the ischemia can cause discomfort or pain in the legs and feet during exercise when the increased muscle activity requires additional oxygen-rich blood. In extreme cases where there is substantial enough blockage that the circulation is severely reduced, a condition referred to as critical limb ischemia can occur. An individual that develops this condition may present with sores that develop on the legs and feet and the individual can experience pain even when resting. Peripheral Artery Disease is diagnosed with the utilization of a few different methods. Tests used in forming a diagnosis could include the angiogram where a contrast dye is introduced into the blood stream and pictures of its flow are taken using x-rays to show narrowed passageways and their location. Another method is the ultrasound which uses high frequency sound waves and echo technology to provide its user with a live picture of the blood movement as it is happening. Comparisons are made of the blood pressure readings that are taken at different locations on the legs and with pressures taken from other body limbs. This is helpful in recognizing potential problem areas. Treatments that are used to manage Peripheral Artery Disease include lifestyle changes such as altering eating and exercise habits and the use of medications to reduce harmfully high levels of cholesterol and triglycerides. Another option is to perform a procedure called the angioplasty. This minimally invasive procedure utilizes a balloon on a catheter tube to reopen narrowed vessels. To perform this process the balloon is fed into the arterial vessel to pre-determined locations where the passageway has been compromised by plaque build-up. The balloon is then inflated to push plaque back against the artery walls and restore blood flow. Surgical options would involve the endarterectomy in which a Doctor surgically opens the artery and removes the plaque, or the Coronary Artery Bypass Graft (CABG) procedure in which blocked locations in the coronary arteries are surgically bypassed using blood vessels removed from other locations in the body to restore blood flow. The chances of acquiring Peripheral Artery Disease increase with age. It is estimated that 8 million Americans are afflicted with it and that up to 20% of those age 70 or greater have the disease.

Did You Know?

There is approximately sixty-thousand miles of blood vessels that run through your body, with enough blood pressure generated within your arteries to send a stream of blood five feet into the air if they are opened.

Postural Orthostatic Tachycardia Syndrome

Postural Orthostatic Tachycardia Syndrome, also referred to as "P.O.T.S.," is an abnormality that affects the body's heart rate and blood pressure when positional changes are made such as the sudden rising to standing position from sitting or laying prone. During these episodes the body fails to make changes to create the proper heart rate and blood pressure adjustments to accommodate for the increased gravitational pull. Postural Orthostatic Tachycardia Syndrome can be induced by various underlying reasons such as abnormalities that are present within the autonomic nervous system, various medical health issues, or diseases that are present within the body. In the normal situation changes such as a minimal heart rate increase, contraction of blood vessels, and a slight rise in the diastolic blood pressure are instituted by the body to compensate for the gravitational change. These changes ensure that there will be an adequate supply of oxygen-rich blood to the vital organs, mainly those that are positioned further from the heart such as the brain. Unfortunately due to conditions that cause low blood pressure or insufficient cardiac perfusion, these changes cannot be made to their full potential. This diminished blood supply creates a response from the body such as a tachycardic heart rate to provide more oxygen-rich blood volume, precipitating various symptoms responses within the body. Some of these responses could include:

* pre-syncopal/syncopal episodes that cause dizziness, lightheadedness, or loss of conciousness
* low blood pressure
* increased heart rate that exceeds 100 beats per minute that is normally 120 or higher
* shortness of breath
* chest pain (angina)
* palpitations
* fatigue/weakness
* the inability to concentrate/difficult memory recall/confusion

Treatment for this condition would center on identifying the underlying reason that is inducing the syndrome. If and when the proper diagnosis is made the appropriate corrective action can be instituted to correct it. Changes in an individuals lifestyle such as regular exercise can help by improving the function of the cardiovascular system and increasing blood flow.

Interesting facts about Postural Orthostatic Tachycardia Syndrome:

* women have up to a 5 time greater chance of being affected.
* the average age of those afflicted is generally between 15 and 50 years.
* genetic factors may have a role in this condition.
* individuals with P.O.T.S. can exert up to 5 times more energy to stand than those who are not afflicted with the syndrome.
* the heart rate can increase by as much as 30 beats per minute or greater, with the increase occurring immediately to within 10 minutes of the positional change.

Prolonged QT Interval

Prolonged QT Syndrome, also called "Long QT Syndrome," is an abnormality in the conduction system that causes a prolonged ventricular repolarization phase of the cardiac cycle. This phenomenon may be congenital, having been genetically introduced into the individual since the earlier stages of development, or it can be a side effect related to the use of drugs such as amiodarone, beta blockers, over the counter antibiotics, or antihistamines. In the normal cardiac cycle the QT Interval time duration will increase or decrease according to the heart rate, becoming longer with slower heart rates and shorter with increased heart rates. The repolarization process associated with this conduction abnormality causes the cells to remain actively recharging within the purkinje fibers for an extended period of time, producing a QT interval that stays prolonged and does not make adjustments or shorten as the heart rate increases and the R-R interval decreases. This action presents cardiac cells that are still recharging within the refractory period that are vulnerable to the electrical impulse stimulation from the oncoming cardiac cycle; creating a situation that has the potential of inducing potentially fatal arrhythmias such as Torsades de Pointes (Polymorphic Ventricular Tachycardia) or Ventricular Tachycardia (Monomorphic Ventricular Tachycardia). This induced premature ventricular response is referred to as "Triggered Automaticity." Under normal circumstances LQTS displays no symptoms, making early diagnosis improbable without the aide of an elective electrocardiogram (EKG) done in conjunction with a stress test. It is under physical or emotional stress that EKG changes become evident or the individual may begin to experience symptoms such as shortness of breath, syncope (fainting), heart palpitations, or the possibility that it may result in cardiac arrest and sudden cardiac death. Long QT Syndrome is present in approximately 1 of every 4000 individuals, but due to its genetic origin in approximately 85% of the patients treated is non-preventable. Treatment can come in a couple of different modalities such as the use of medication and lifestyle changes. These changes would reduce the possibility of recurrences but due to its nature of going unnoticed until there is an event the most prevalent first response treatment would be symptom management and the termination of the triggered arrhythmia. For long term treatment an Implantable Cardioverter Defibrillator or ICD would be surgically implanted to terminate any new arrhythmia episodes triggered by this condition.

The normal time duration measurements for the QT Interval are calculated from the beginning of the Q-wave to the end of the T-wave, which is the beginning of the ventricular depolarization to the completion of the ventricular repolarization. These time duration measurements would normally be 0.36-0.44 one hundredths of a second for men and 0.36-0.45 one hundredths of a second for women. The general rule is that the QT interval should measure out to be one-half or less than the total of the R-R interval, which is the heart's intrinsic ventricular pulse interval or ventricular depolarizations. For example an R-R measurement of 0.86 one hundredths of a second should have a QT interval at or less than 0.44 one hundredths of a second, providing all other measurements of the cardiac within normal limits (PR 0.12-0.20 one hundredths of a second, QRS 0.06-0.10 one hundredths of a second) and the heart rate is between 60-100 beats per minute.

Pulseless Electrical Activity

Pulseless Electrical Activity, which is also referred to as "PEA" or "Electromechanical Dissociation," is the absence of a pulse when there is electrical activity present within the myocardium. This is a situation where the heart muscle is unable to circulate the blood through the body despite the presence of impulses delivered through the conduction system. PEA is a condition that is caused by an abnormality that originates from a secondary source that compromises the mechanical mechanisms of the myocardium or the circulatory system. Possible occurrences that may precipitate this situation might include a metabolic imbalance, hypovolemia which is low blood circulating volume, or an issue such as a tension pneumothorax. Some additional factors that may possess the potential to induce this condition would include:

* hyperkalemia / hypokalemia – abnormal potassium levels being too high or too low that are present within the body.
* hypoxia – low blood oxygen levels that may be precipitated by air exchange within the lungs or a condition referred to as ischemia; although this term commonly involves the amount of oxygen within the blood, not a diminished supply of blood.
* hypothermia – below normal core body temperatures. Hypothermia occurs when the body's core temperature drops below 95F. This presents a condition that adversely affects the body's functions and ability to metabolize properly.
* cardiac tamponade – constrictive pressure to the exterior of the heart that inhibits its ability to expand. This situation can be induced by a condition called pericarditis in which the pericardium becomes inflamed or there is a fluid build-up from an outside source or a traumatic injury.
* toxicity – body's reaction to chemical or drug poisoning.
* thrombus – blockage in a major artery caused by a blood clot that interrupts the blood flow to the heart or lungs.

The main objective in dealing with this event is to protect and sustain the life of the individual with "Basic Life Support" or "Advanced Cardiac Life Support" while the factors that precipitated the PEA are established. This is accomplished by utilizing lab tests and other diagnostic resources that can be executed and resulted in a rapid manner. Some of these would include:

* lab work to check electrolyte levels present within the body.
* arterial blood gases to check Ph level, oxygen saturation level, etc. This test evaluates blood from an artery that has been freshly re-oxygenated to check gas levels.
* electrocardiogram (EKG) - this test will show abnormal cardiac rhythm activity such as arrhythmias like heart block, peaked t-waves that may indicate hyperkalemia, flattened T-waves that may be an indicator of hypokalemia, prolonged QT interval that can indicate a drug overdose, a heart attack can be identified through the observation of the ST segment elevation or depression with regard to its relationship to the isoelectric baseline.

Until the underlying cause of the condition has been established, this emergency situation is treated using the same algorhythm as asystole. Once the cause of this condition has been established the proper medical intervention can be employed to correct the abnormality. Some of these may include the introduction of certain drugs such as epinephrine or atropine to stimulate cardiac activity, thoracostomy which is the surgical incision into to chest wall to provide an opening to allow drainage, pericardiocentesis to relieve external pressure on the heart precipitated by tamponade (emergent care would require the removal of excess pericardial fluid with a syringe), the replacement and restoration of metabolic balance of the body's chemistry, increase the circulating blood volume through volume infusion, the utilization of thrombolytics such as Tissue Plasminogen Activator (tPA) to dissolve clots and re-establish blood flow and the administration of oxygen in cases when hypoxia is present.

Defibrillation, or the intentional delivery and introduction of an electric shock to depolariza the hearts conduction system would produce no results due to the fact that the abnormality originates within the mechanical workings of the myocardium; not the conduction system

Did You Know?

The heart begins to develop as soon as the embryo starts developing inside the mother's womb, with the chamber walls taking form as soon as 6 weeks old. It grows so fast that there is not enough room for it to become longer, so it doubles back and twists, taking on the appearance we recognize. There is a layer of tissue called the septum that grows down the middle at this point dividing the right and left sides. A male heart weighs approximately 11 ounces, compared to 9 ounces of the heart of a woman. It is thicker at the top, is approximately five inches long, three & one-half inches wide, and two & one-half inches thick. It is located in the middle of the chest in a horizontal position with the lower part of the heart pointed to the left side of the body. The left ventricle muscle walls are three times as thick as the right ventricular walls to provide a better catalyst to move the blood throughout the entire body. It beats 70 times per minute (normal heart rate), or 100,800 times a day, 90 times a minute for a 7-year-old child, and 120 times a minute for an infant. The heart pumps five quarts of blood through its chambers every 60 seconds.

It is estimated by the FDA that in the year two-thousand and six twenty-five percent of the heart attacks (roughly around one-hundred and seventy-five thousand) were going to be "silent," These episodes would display minimal symptoms or be mistaken to indicate muscle soreness. Unfortunately, heart attacks evolve over time and cause more myocardial damage the longer they go untreated and these unrealized cases take more time to receive medical attention.

Premature Ventricular Contraction

Premature Ventricular Contractions, also referred to as "PVC's" or "Premature Ventricular Complexes," are early ventricular depolarizations that are caused by abnormal ectopic impulses that are created within the ventricles. The premature firing of the ventricles initiated by these cells causes them to contract earlier than they would have if they followed the depolarization sequence of the impulse normally initiated by the sinoatrial node. They are wide and bizarre in appearance and can be seen in many different forms and configurations. Some of these would include:

* uni-focal – meaning that they originate from the same focal point in the ventricles. All PVC's will possess the same appearance.
* multi-focal – they were formed from different locations within the ventricles, giving them different appearances.
* inter-polar – occurring between two ventricular depolarizations without changing the R-R intervals of the rhythm
* compensatory pause – this is a pause that accompanies the PVC that fills in the time duration until the next normal beat was scheduled to occur. This allows the next intrinsic beat to fall in line to keep rhythm.
* group beating – this is when there are multiple PVC's happening in a row which is also referred to as runs or salvo's. They can be observed in the following patterns:
    * couplet – two occurring in a row.
    * triplet – three occurring in a row.
    * more than three in a row is considered non-sustained v-tach if under thirty seconds in duration (sustained v-tach is measured by a time duration of thirty seconds or greater).
    * bi geminy - every other beat.
    * tri geminy – every third beat.
    * quadra geminy – every fourth beat.

Although the primary cause of this ectopy may sometimes remain unknown, there are known conditions or factors that contribute to their presence. Some of these may include fatigue, stress, ischemia, caffeine, alcohol, nicotine, or an imbalance in the body's electrolytes. Premature ventricular contractions are not normally harmful, but can be of a more serious nature in situations where there is already the existence of heart disease. However, due to shortened ventricular refilling time due to the quick depolarization induced by the PVC's occurring in a run, cardiac output can be adversely affected. This situation can induce negative physiological issues such as pre-syncope, syncope, or shortness of breath that can happen due to an insufficient volume of oxygen-rich blood to the body. Interventions can include antiarrhythmic drugs, the introduction of oxygen, or the replacement of depleted electrolytes in patients with below normal levels.

Sick Sinus Syndrome

Sick Sinus Syndrome, which is also referred to as "Brady-Tachy Syndrome" and "Sinus Node Dysfunction," is an arrhythmia that is induced by the sinoatrial node. This arrhythmia is considered rare and is seldom seen in individuals less than fifty years of age and is not normally recognizable until it is in its advanced stages and may present with or without identifiable symptoms. In the normal heart rhythm the sinoatrial node generates and delivers electrical impulses into the conduction pathway in a synchronized and regulated fashion. Sick Sinus Syndrome produces electrical impulses in an erratic manner that can increase and decrease the heart rate without warning. These episodes can be sustained for long periods of time or paroxysmal in nature and happening in short bursts. This behavior can induce heart arrhythmias such as sinoatrial exit block, supraventricular tachycardia and sinus bradycardia. In extreme cases this abnormality can also progress to total sinus node failure that results in sinus arrest. Sick Sinus Syndrome's appearance on the heart monitor displays frequent and abrupt heart rate changes from tachycardic heart rates that reach greater than 100 beats per minute to bradycardic heart rates that drop too low (generally below 60 beats per minute). These changes will commonly be accompanied by long pauses that occur as the heart rate changes from tachycardic to bradycardic. Causes for this conduction disturbance would include disease of the sinoatrial node, reaction to medications such as beta-blockers, digitalis, hyperkalemia, damage to the myocardium caused by a prior heart attack, hypothyroidism, cardiac ischemia, hypoglycemia, cardiomyopathy, or sleep apnea. Sick Sinus Syndrome is diagnosed mainly through the use of an electrocardiogram (EKG) which display's erratic changes in the hearts electrical conduction behavior. Possible side effects experienced due to this abnormality would be induced mainly by the compromised cardiac output created by the increased or decreased heart rates encountered. The body's symptomatic responses can include angina, shortness of breath, dizziness, fatigue, syncope, or palpitations. Treatment for this condition would be to manage and regulate the hearts conduction system with the utilization of a pacemaker which would be employed to resynchronize the heart's chamber contractions. The implementation of the pacemaker is also necessary if medications are to be utilized due to treat bradycardic and tachycardic heart rates. This is an important safeguard to prevent situations such as extreme bradycardia when using medication to regain and control rapid heart rates. The chances of becoming afflicted with Sick Sinus Syndrome increase with age and it is estimated that it is present in approximately 2 to 3 individuals in every 10,000.

Did You Know?

Statistics show that women have a greater risk of dying from some form of cardiac related disease. In the year 2003 there were over two-hundred thirty thousand women who died from cardiovascular disease. Over eighty thousand died from a heart attack, with an average age of seventy years old. Of the heart attack survivors there were more than thirty five percent that died within the first year of the initial onset.

Sinus Bradycardia

In this rhythm all of the electrical impulses originate at and are delivered from the site of the sinoatrial node which is located in the upper right atrium. These electrical impulses then follow a prescribed conduction pathway through the heart muscle initiating a firing rate under sixty beats per minute. There are three preset cardiac cycle time duration guideline measurements that it will need to be within. The first measurement will be from the start of the atrial depolarization (P-wave) to the start of the ventricular depolarization (QRS) and should be within the time frame of 0.12 to 0.20 one hundredths of a second. This is represented by the PR interval on the electrocardiogram tracing. The next measurement is the time duration from the start of the ventricular depolarization to the completion of the ventricular depolarization. This is calculated from the beginning of the Q-wave to the J-point and should measure to be within the prescribed guidelines of 0.06 to 0.10 one hundredths of a second. The QT interval represents the entire time duration from the start of the ventricular depolarization to the completion of the ventricular repolarization phase. The guidelines for this entire process should be within the set parameters of 0.36 to 0.44 one hundredths of a second. The rhythm will be regular with one atrial depolarization (P-wave) for each ventricular depolarization (QRS complex), with one upright and rounded T-wave which represents the ventricular repolarization. Sinus bradycardia can be perpetuated by various factors such as the imbalance of electrolytes, drug reaction to medications such as digitalis and beta blockers, hypoglycemia, myocardial infarction, hypothyroidism, sleep apnea and infection; however, this may also be a bodily response to increased vagal tone through stimulation of the vagus nerve caused by vomiting, straining during a bowel movement or procedures such as intubation. This rhythm may also be seen in athletes and individuals who maintain a healthy cardiovascular system through exercise routines or may be associated with a condition that is referred to as "Brady/Tachy Syndrome" in which diseased tissue within the sinoatrial node disrupts the heart's production of the normal electrical impulse. Sinus Bradycardia does not normally produce symptoms unless the heart rate drops to a very low level at which point the individual may experience fatigue, palpitations, syncope, shortness of breath, lightheadedness and chest pain due to the insufficient supply of oxygen-rich blood. At these occurrences mechanisms located further along the conduction system such as the AV junction or purkinje fibers may take control through automaticity. This will occur when the heart rate of the sinoatrial node falls below the intrinsic rates of the AV junction which is 40-60 beats per minute or that of the purkinje fiber network that fires at a rate of 20-40 beats per minute. Sinus bradycardia is diagnosed through the utilization of an electrocardiogram (EKG), but other factors related to the slow rhythm are visible upon presentation. Some of these would include peripheral edema, inability to orientate, confusion, dyspnea and cyanosis from poor circulation. Treatment in symptomatic cases would be to provide therapy to increase the heart rate and eliminate the low cardiac output. This is accomplished medicinally with drugs such as atropine, or mechanically to manage the heart's conduction with artificial stimulation the heart's conduction system with the introduction of a surgically implanted pacemaker.

Sinus Tachycardia

Sinus tachycardia is classified as a rhythm that possesses a heart rate between 100 and 160 beats per minute in which the conduction is controlled by the delivery of electrical impulses that originate in the sinoatrial node. Observing the fact that the classification of tachycardic heart rates between that of infants (110-150) and adults are different, this article will pertain to adults. As it is with the normal sinus rhythm of the conduction system the electrical impulse originates and is delivered by the sinoatrial node which is a cluster of cardiac cells that is located in the upper right atrium. These cells possess automacity which gives them the ability to create their own electrical impulse without outside stimulation. This electrical impulse then follows a prescribed impulse pathway through the heart muscle initiating a synchronized sequence of muscular contractions to move the blood from the hearts chambers to the lungs and body. The time duration measurements will to be within the normal limits of PR 0.12-0.20 one hundredths of a second from when the atrial depolarization occurs to the beginning of the ventricular depolarization. This phase is represented by the QRS in the EKG tracing and should measure to be within 0.06-0.10 one hundredths of a second. The QT interval should be within a time frame of 0.36-0.44 one hundredths of a second for ventricles to start and complete their depolarization and repolarization phases. The rhythm will be regular with one atrial depolarization which is displayed on the tracing as an upright and rounded P-wave for each ventricular depolarization as represented by the QRS complex. The P-wave will in most cases remain visible during this fast rhythm, but in some instances such as those that involve rapid rates the P-wave may be within the preceding T-wave; causing it to be hidden. This increase in the heart rate can be caused by outside factors such as caffeine or nicotine, it may be the body's natural response to different variables such as fever, stress, pain and dehydration. This is referred to as the body's "fight or flight" response which is a protective mechanism that prepares the body to be able to deliver an adequate response to the perceived occurrence of imminent physical or emotional duress. At these times the sympathetic nervous system releases adrenalin and increases the glucose and oxygen (through blood supply volume) to provide a heightened strength. Another factor could be the presence of coronary artery disease (CAD). This disease causes the arterial lumens (passageways) to become narrowed from the accumulation of plaque and fatty deposits, necessitating the heart muscle to work harder to provide more blood to compensate for the diminished volume. This situation would also induce a tachycardic heart rate. The sinus tachycardia rhythm does not normally produce symptoms, however, treatment would be necessary for sustained episodes that would involve a decreased cardiac output due to shortened chamber refilling time. This would result in an insufficient blood supply to satisfy the body's needs, creating a condition called ischemia. Other forms of this rhythm can be seen in a syndrome called POTS (Postural Orthostatic Tachycardia Syndrome) or an abnormal autonomic condition referred to as Inappropriate Sinus Tachycardia. Treatment if necessary would incorporate the use of medications or therapy to counter the initial factors responsible for inducing the increased automaticity.

Stroke

Also referred to as a "Cerebrovascular Accident" or "CVA," the stroke is a medical emergency that occurs due to a lack of oxygen-rich blood flow to the brain. This event may be precipitated by conditions such as ischemia induced by Coronary Artery Disease, or it can be the result of a blockage in a cerebral artery by a thrombus or embolism. Depending upon the amount of time that the cells are deprived of oxygen-rich blood, irreversible damage can be inflicted to the brain as a result of necrosis (cell death). This occurs when brain cells die leaving residual side affects to the parts of the body that are controlled by the part of the brain suffering the injury. There are two different types of strokes, they are the Hemorrhagic and the Ischemic; here is a brief description of each.

Hemorrhagic – this form of stroke is the result of hemorrhaging within or around the brain. The different forms of bleeding associated with a hemorrhagic stroke are:

* cerebral hemorrhage - which is bleeding inside of the brain from ruptured smaller arteries causing blood to build-up. This form of hemorrhage is often associated with high blood pressure and is responsible for about nine or ten percent of strokes.
* subarachnoid hemorrhage – this type of hemorrhage happens when there is bleeding occurring around the brain. Commonly precipitated by a head injury or a ruptured aneurysm. This form of hemorrhage happens more often in younger individuals and is responsible for three or four percent of occurring strokes.

Ischemic – the ischemic stroke occurs more often and accounts for more than 85% of all strokes. Ischemia is a condition that is caused by the lack of oxygen-rich blood that is supplied to the body's cells. Cerebral Ischemia is caused by a cerebral thrombus or an embolism that has traveled through the blood stream and become lodged in an artery that supplies oxygen-rich blood to the brain. There are two cardiovascular events that could precipitate this event, they include:

* embolic stroke – caused by a blockage in a artery that feeds the brain. This occurs when an embolism has traveled through the cardiovascular system and becomes lodged in a cerebral artery. This form of stroke occurs suddenly and without warning.
* thrombotic stroke – is precipitated by the formation of a clot within a cerebral artery that has been compromised by plaque build-up, causing the opening to narrow. This is a problem that happens gradual and over time that causes a stoppage or serious reduction in the volume of oxygen-rich blood that is supplied to the brain.

To prevent extensive residual physical damage when a stroke occurs it is necessary to receive immediate treatment to restore the blood supply to the brain. This is a process called "reperfusion." To accomplish this it is first necessary to evaluate and decide exactly what form of stroke is occurring. Once this has been determined, the following methods are utilized to correct the situation.

* for a situation involving a blockage due to a clot, the thrombolitic drug TPA which stands for "Tissue Plasminogen Activator" is used. This drug dissolves the blood clots and clears the opening in the blood vessel for renewed blood flow. Although this is effective for the thrombolitic stroke, it can be dangerous if used in treating cerebral or subacarcnoid strokes that involve ruptures where there is bleeding in or around the brain by precipitating uncontrolled excessive bleeding.

* in the instance where there is an ischemic problem due to a compromised blood vessel lumen due to a plaque build-up such as those associated with "coronary artery disease," a couple of procedures are utilized.
* the first is referred to as a "carotid endarterectomy." This procedure involves the opening of the carotid artery to remove fatty deposits and re-establish blood flow through the arterial passageway. An incision is made in the neck to expose the artery and the blood is re-routed at the point before and beyond the narrowed section. The abnormal section of the artery is then repaired or a graft is surgically placed to correct the problem.
* the second is a called a "carotid angioplasty." This procedure utilizes a balloon tip catheter to reshape the blood vessel by expanding the balloon and compressing the plaque back against the artery wall. If it is necessary after the process of re-opening the vessel stents may be introduced within the vessel to maintain its integrity and keep the passage open.

The stroke is one of the leading causes of disabilities that are classified as long term. It is estimated that of the patients that experience strokes, 40% to 70% will recover to a point that they are able to function on their own and some will suffer temporary disabilities; however:

* 20% will need institutional care
* 22% to 25% will die within the first year following the episode
* 14% will experience another episode within a year
* 15% to 30% will be disabled permanently with disabilities that may include one or more the following:
  * vision impairment or loss
  * speech impairment
  * swallowing difficulties
  * bowel and bladder retention abilities
  * memory retention / confusion
  * lack of strength/weakness
* It is estimated that one out of every twenty people may experience a stroke before they reach seventy years of age.

There are many risk factors that may contribute to the onset of a stroke. Some of these may be corrected with lifestyle changes, however, those that are precipitated by factors such as age, family history and surgical procedures cannot. Factors within ones ability to control through everyday changes can include:

* anti-coagulation of individuals that have certain heart arrhythmias such as Atrial Fibrillation and Atrial Flutter which precipitate blood pooling carry the increased risk for the formation of clots that may travel through the blood stream precipitating an embolic stroke.
* monitoring of blood pressure for high risk individuals. Uncontrolled high blood pressure presents a four to six percent greater risk of having a stroke.
* cessation of cigarette smoking.
* restraining from the excessive use of alcohol and drugs.
* control eating and increase exercise habits to avoid obesity.
* controlling diabetes.

Supraventricular Tachycardia

Supraventricular Tachycardia which is also called "SVT," is a heart arrhythmia caused by a conduction disturbance that is induced from above the ventricles that increases the heart rate between than 160 and 250 beats per minute. This arrhythmia can be sustained in nature and lasting for an extended period of time, or paroxysmal and happening in short bursts that can be seen to start and stop. This anomaly is referred to as PSVT or Paroxysmal Supraventricular Tachycardia. There are various atrial induced arrhythmias that can be categorized into this classification, a few of these would include:

* Rapid Atrial Fibrillation – this arrhythmia is the result of disorganized electrical activity created by multiple electrical impulses firing within the atria without order. This rhythm is irregular and will possess no discernible p-wave.
* Rapid Atrial Flutter possesses F-waves that are saw tooth in appearance. This arrhythmia can be seen as variable conduction which makes the R-R intervals irregular, or when the heart rate that is at or exceeds 160 beats per minute will commonly be a 2:1 conduction with regular R-R intervals.
* Multi-Focal Atrial Tachycardia – which is also referred to as MAT, is a rhythm that possesses a minimum of three different p-wave morphologies that will display different time interval measurements in the PR and R-R segments. If not carefully evaluated, MAT can be mistaken for atrial fibrillation.
* Inappropriate Sinus Tachycardia – this is caused by an abnormality within the sinus node that can induce a rapid heart rate that can occur both during exercise and at rest.
* Wolf Parkinson White syndrome which is also called "WPW," is identifiable by the presence of an accessory conduction pathway that allows the electrical impulse to travel unregulated from the atria to the ventricles. This arrhythmia can perpetuate extremely fast and irregular heart rhythms.
* Re-Entrant Tachycardia is a circuit re-entry arrhythmia caused by abnormalities within the AV node in the form of an extra pathway. Under certain conditions, this allows a rapid heart rate through an anomaly called pre-excitation syndrome. Factors that may induce this arrhythmia can include stress, exercise, caffeine, certain medications or illicit drug use.

Medicinal treatments utilized to control this arrhythmia would include medications such as calcium channel blockers, beta blockers. Other medical treatment for correction of this anomaly could include a procedure called radiofrequency ablation. This minimally invasive procedure performed by an Electrophysiologist and is performed to neutralize areas in the AV junction that are causing abnormal behavior. This is accomplished by employing radiofrequency energy through an electrode at the end of a catheter to destroy problematic tissue or closing unregulated pathways. Radiofrequency catheter ablation first came about in the 1980's but wasn't used with regularity until the 1990's.

Torsades de Pointes

Torsades de Pointes, also called "Polymorphic Ventricular Tachycardia," is a variant of the ventricular tachycardia arrhythmia that can sustain rates between 150-250 beats per minute. This arrhythmia was first diagnosed in the year 1966 by Dr. Francois Dessertenne and acquired its name which in French means "twisting of the points" in reference to its unique morphology that due to amplitude changes makes it appear to be twisting around the isoelectric baseline. Torsades de Pointes is believed to be induced by different factors such as hypoxia, hypokalemia, hypomagnesemia, heart failure, bradycardia, age, hypertrophy of the left ventricle, or class 1 and class 3 antiarrhythmic drugs. However, the main reason for the onset of this arrhythmia is believed to be Long QT Syndrome which can be of a congenital nature or acquired from the aforementioned drugs. This syndrome which is also referred to as "Prolonged QT Interval" produces an extended ventricular repolarization phase of the cardiac cycle that present cardiac cells still in flux as the new impulse enters. This vulnerable state provides the opportunity for the impulse to prematurely stimulate the still charging cardiac cells, inducing the Torsades de Pointes arrhythmia through "triggered automaticity." The normal QT interval time duration is up to 0.43 hundredths of a second with a maximum time duration a little bit longer for women at 0.45 one hundredths of a second as compared to a maximum duration of 0.44 hundredths of a second for men. Individuals stricken with this arrhythmia will present with various symptoms caused by the drop in arterial blood pressure and diminished oxygen-rich blood perfusion. Some of these symptoms could include dizziness, syncope and shortness of breath. This arrhythmia if left untreated can deteriorate to the arrhythmia called Ventricular Fibrillation, presenting a worsening the prognosis for recovery or resulting in sudden cardiac death. One of the treatments utilized in terminating this arrhythmia is defibrillation. This life saving procedure delivers an electrical shock to depolarize the myocardium and create a momentary interruption in the conduction system. This provides the opportunity for the sinoatrial node to regain control when the heart restarts. Drug therapy would include an infusion of magnesium sulfate and antiarrhythmia drugs such as dilantin and lidocaine.

Lidocaine (Parenteral) Antidysrhythmic Class 1B

Developed in the year 1943 by a Swedish scientist named Nils Lofgren, Lidocaine is used to stabilize the cardiac membrane in arrhythmias such as ventricular tachycardia by increasing the electrical stimulation threshold of the ventricles/His purkinje fibers. This drug accomplishes this by inducing a decrease in the hearts automaticity by blocking the sodium channels that feed the cells of the electrical conduction system of the heart. This drug when employed in the cardiac setting is introduced into the system during cardiac arrhythmias or cardiac arrest through intravenous injection to aid in breaking the potentially fatal rhythm to regain control of the hearts conduction system. Lidocaine needs to be monitored on telemetry for increased ventricular ectopy and any radical changes that may occur within the conduction cycle intervals.

Transient Ischemic Attack

"Transient Ischemic Attacks" which are also referred to as a "TIA's" or "Mini Strokes," are transient ischemic neurologic events that present for only short time durations. They will commonly last for a total of eight to fourteen minutes, but must be under 24 hours to be classified as a Transient Ischemic Attack. Episodes that last for more than 24 hours are classified as a "Cerebrovascular Accident" (CVA) or "Stroke." As the word ischemia suggests, the TIA is an event that occurs due to the lack of oxygen-rich blood flow to the brain which causes the tissue to infarct and eventually die through necrosis. Although there are different conditions that can induce the mini stroke, one of the factors that can precipitate this event is a condition known as Coronary Artery Disease. In this disease a build-up of fatty deposits and plaque within the arterial lumens causes a constriction that diminishes the volume of blood flow to the body's tissue resulting in a condition called ischemia. Other factors that may increase the risk of experiencing a TIA can include diabetes, uncontrolled high blood pressure, elevated cholesterol levels, the interrupted flow of blood caused by a clot or embolism, or an intracranial bleed from an artery or vein. Although they are short in duration, the signs are the same as those produced by a regular stroke and can last for an entire day. Some of the symptoms would include confusion, vision damage or loss, speech impairment, numbness and loss of feeling on affected parts of the body and facial droop.

To diagnose this medical emergency and to locate and evaluate possible factors that may have caused the event, the following methods are a few of those utilized:

* CT Scan – this test can provide immediate evidence of hemorrhage or blockage that involves a cerebral artery.
* Carotid Duplex Scan – in the Carotid Scan the arteries that feed the brain its blood supply are examined for potential blockages from conditions such as ischemia and to wht degree.
* Cerebral Angiogram – used in conjunction with a contrast medium (dye), the angiogram uses an x-ray to study the blood flow within the cerebral circulatory system. This test will show blockages or disturbances in the blood flow.
* Magnetic Response Imaging (MRI) – when performed on the head will display areas involving infarct, ischemia, or brain swelling.

Emergency treatment would be to resolve blockages and re-establish blood flow. To accomplish this goal a drug referred to as "Tissue Plasminogen Activator" (TPA) or "Clot Buster" is introduced into the system. This antithrombolytic chemical dissolves the clots that are causing the blockage to induce revascularization, but is most effective if used within the initial 3 hours of the onset to receive its optimum benefit. Other procedures that can reduce the recurrence of these events would include the minimally invasive procedure referred to as the Percutaneous Transluminal Coronary Angioplasty or "PTCA." The angioplasty is used to open up narrowed passageways (lumens) within the arteries to re-establish blood flow. During this procedure plaque and fatty deposits are pushed back against the artery walls with an inflatable balloon to reestablish blood flow through the vessels. This procedure is very useful in resolving issues where there is the presence of coronary artery disease.

The endarterectomy is a surgical procedure where the carotid arteries are surgically opened and plaque and fatty build-up is removed by the surgeon. Medicinal treatment would include anticoagulation therapy. In this maintenance program drugs such as heparin and coumadin are administered to prevent coagulation and clot formation. Although these drugs can not dissolve clots that are already present, they are effective in inhibiting necessary clotting factors such as thrombin and vitamin K.

It estimated that one in every three individuals who experience a transient ischemic attack will suffer a cerebrovascular accident (CVA) and that as high as fifty percent of these will occur within a year from the initial episode. It is also estimated that men have an increased risk of suffering a transient ischemic attack over that of a woman and that black men have an increased risk over that of white men. Studies show that occurrences in the age group from 45-50 years of age is 16 per 100,000 individuals with an increase in 4-8 cases per 1,000 individuals between the ages of 50-60 years.

Did You Know?

Incidents that involved the restenosis of blood vessels that have been re-opened through the angioplasty procedure has decreased due the discovery and use of medicated stents named "Cyphers." These stents were developed by the Johnson & Johnson Company and were approved by the Food & Drug Administration (FDA) in the year 2003. These stents release a medication that inhibits the excessive cell growth that can occur after these procedures.

The Percutaneous Transluminal Coronary Angioplasty procedure or "PTCA" can also be done with the employment of a laser. As with the balloon procedure the catheter tube is inserted into the artery in the arm or leg and guided to the site of the blockage. At this point the laser at the tip of the catheter tube burns away plaque build-up.

There is an estimated risk that out of every one thousand angioplasty procedures performed one to four due to various reasons could result in death. There is also up to a five percent chance that there will be emergency open heart intervention and that as many as three percent of these procedures can induce a heart attack.

The U-Wave

The U-Wave, which is also called an "afterdepolarization," is an abnormality in the hearts conduction system that was identified and named by William Einthoven in the year 1903. This phenomenon occurs within the ventricular repolarization phase of the cardiac cycle that has been thought to be an anomaly that may be induced by a couple of different factors. One possibility is a delayed cellular repolarization of the purkinje fibers that are located beneath the endocardium of the ventricles that conduct, regulate and transmit the electrical impulse to induce the ventricular depolarization. Another possibility is that it is thought to be the presence of ischemia. It has been observed that prominent U-waves that were visible before the Percutaneous Transluminal Coronary Angioplasty (PTCA) procedure were no longer present after the procedure and revascularization has been accomplished. This abnormality presents itself on the electrocardiogram (EKG) as a rolling swell in the isoelectric baseline that immediately follows the T-wave. Its deflection is normally seen to be in the same direction as the T-wave, however, in cases where there is a history of prior heart attack, exercise induced hypoxia, or the presence of myocardial ischemia; it can be in the opposite direction of the T-wave. The U-wave may also be caused by various other factors, some of these may include:

* hypokalemia – low potassium level in the blood, of which 95% of the body's total is in the cells
* antidysrhythmic medications such as quinidine, digitalis, or epinephrine
* hyperthyroidism – overactive thyroid gland
* hypercalcemia – elevated calcium levels
* Long QT Syndrome which is a condition that may be congenital or drug induced that causes a lengthened or prolonged ventricular repolarization phase of the cardiac cycle.
* Mitral Valve Prolapse (MVP) – an improperly positioned and dysfunctional mitral valve that does not seal properly. This condition allows regurgitation back into the atria resulting in a weakened perfusion.

The repolarizing cells as indicated by the presence of the U-wave have a side effect that is referred to as "triggered automaticity." This indicates that there is a potentially dangerous situation where there is the potential for abnormal cardiac conduction behavior. Triggered automaticity occurs when vulnerable cardiac cells still actively repolarizing within the relative or supernormal refractory periods from the previous cardiac cycle become stimulated by electrical impulses of enough intensity entering from the new cardiac cycle. The combining of the new electrical impulses into the repolarizing cells can prematurely stimulate a ventricular response and induce a fatal arrhythmia such as Torsade de Pointes or Ventricular Tachycardia.

Did You Know?

Betapace (Sotalol) **Antidysrhythmic** **Class II**

This drug extends the time duration of the ventricular repolarization period and prolongs the QT interval. It is also used in treatment of ventricular dysrhythmias. Betapace needs to be monitored for EKG rhythm changes related to the ventricles such as Torsades de Pointes, the precipitation of heart blocks, or a bradycardic heart rate.

Heart Rate Calculation Methods

1) Count the R-R's in a six second strip, and multiply the number by ten.
2) Count the number of 0.20 second squares between an R-R interval, then divide the number into 300.
3) Count the number of 0.04 second boxes within one R-R interval and use table below to find heart rate.

# of 0.04 boxes	Heart Rate	# of 0.04 boxes	Heart Rate
5	300	26	58
6	250	27	56
7	214	28	54
8	188	29	52
9	167	30	50
10	150	31	48
11	136	32	47
12	125	33	45
13	115	34	44
14	107	35	43
15	100	36	42
16	94	37	41
17	88	38	40
18	84	39	39
19	79	40	38
20	75	41	37
21	72	42	36
22	68	43	35
23	65	44	34
24	63	45	33
25	60	47	32

Note:
Telemetry computers can present incorrect heart rate readings due to the incorrect counting of elevated T-waves as QRS complexes. Questionable heart rates should always be re-counted manually.

Ventricular Fibrillation

Ventricular Fibrillation which is also called "V-Fib," is a potentially fatal arrhythmia that is initiated by disorganized electrical activity within the heart that perpetuates a quivering motion of the heart's chambers in place of the myocardium's natural contractions. This fibrillating motion that is caused by the manner in which the ventricles conduct the electrical impulses diminishes arterial pressure and incapacitates the heart's chambers to a degree that they can not circulate their fill of blood. Ventricular fibrillation can be induced by different factors such as an imbalance in the body's electrolytes, cardiomyopathy, drug overdose, myocardial infarction, Myocarditis, or a deteriorated state of Ventricular Tachycardia or Torsades de Pointes, but heart disease and ischemia are found to be prevalent manifestations in many cases. Another anomaly within the conduction system that can initiate this arrhythmia through triggered automaticity is Long QT Syndrome. This abnormality which is also referred to as "Prolonged QT Interval" is an extended repolarization phase of the ventricles that presents vulnerable cardiac cells that are still in a state of flux susceptible to premature stimulation by incoming electrical impulses. This action induces a circuit re-entry similar to that seen in the looping conduction of atrial flutter. Ventricular Fibrillation can also be idiopathic in nature with no presentable cause for its occurrence. During these episodes the time duration of the event is of the utmost importance. Cardiac output decreases to zero, causing ischemia and necrosis to occur due to the lack of oxygen-rich blood flow. This event cause irreversible damage to the bodily organs and tissue, or in many cases results in "Sudden Cardiac Death" in a matter of minutes. Diagnosis of this arrhythmia is made by examining a tracing from an electrocardiogram or through the monitoring of cardiac telemetry. Individuals will experience shortness of breath, palpitations, chest pain, fatigue and weakness prior to losing consciousness. Although its onset is commonly sudden and without warning, there are abnormalities that predispose individuals to having a higher risk of experiencing the arrhythmia. Some of these could include history of R on T activity, the presence of a U-wave (afterdepolarizations), Long QT Syndrome, ischemic heart disease and a history of recurrent episodes of ventricular tachycardia. Treatment for V-Fib would be to terminate the arrhythmia and to institute a maintenance program to prevent future episodes. To accomplish this on an emergent basis, defibrillation is employed to temporarily depolarize the hearts conduction system. This interruption will give the hearts sinoatrial node the opportunity to regain control and restore sinus rhythm. In the hospital, staff would utilize defibrillation along with antiarrhythmic drugs such as amiodarone and lidocaine to terminate the arrhythmia. Long term treatment would constitute the introduction of antiarrhythmic medications in maintenance doses and the implantation of an Automated Implantable Cardioverter Defibrillator (AICD) to monitor the hearts conduction rhythm and terminate any arrhythmias that occur. Though there are physiological means that induce this arrhythmia that even today are not fully understood, the concept and study of this arrhythmia started in the year 1887 when a Physiologist and Professor of Physiology at Aberdeen University named John MacWilliam provided a description of the ventricular fibrillation arrhythmia that is still applicable today. His description is as follows:

"The ventricular muscle is thrown into a state of irregular arrhythmic contraction, while there is a great fall in the arterial blood pressure, the ventricles become dilated with blood as the rapid quivering movement is insufficient to expel their contents, the muscular action partakes of the nature of a rapid uncoordinated twitching of the muscular tissue. The cardiac pump is thrown out of gear and the last of its vital energy is dissipated in the violent and the prolonged turmoil of fruitless activity in the ventricular walls." Another important finding regarding this arrhythmia came about in the year 1922 when an electrocardiogram that was produced by Bender and Kerr displayed the change that occurred when a ventricular tachycardia rhythm deteriorated into the ventricular fibrillation arrhythmia.

Did You Know?

Amiodarone (Cordarone) Antidysrhythmic Class 3

This class 3 medication extends the time duration of the refractory period, prolongs the PR and QRS intervals and slows the sinus rate. Amiodarone was developed in the year1961 in Belgium by Dr. Bramah Singh to be used primarily as a treatment for angina. During it's use as an anti-angina medication, amiodarone was found by Dr Singh to be equally effective in the suppression of both ventricular and supraventricular arrhythmias due to its prolongation of the heart's repolarization cycle and QT interval, slowing of the electrical impulses as they travel through the heart and it's effectiveness in decreasing the heart's automaticity. Its use has produced good success among his patients. With this news an Argentinean physician named Dr. Mauricio Rosenbaum utilized amiodarone on his patients while keeping reference records of his successes. These notes eventually made their way to physicians in the U.S .in the 1970's, though due to non-acceptance by the FDA, their supplies had to be obtained in Canada. Approval by the FDA came about in the later part of 1985, classifying it as a Class lll Antidysrhythmic. Amiodarone has multiple side effects that in some instances can prove fatal, but due to its effectiveness and success in converting arrhythmias where other antidysrhythmics have failed to produce results, it is commonly used.

Lidocaine (Parenteral) Antidysrhythmic Class 1B

Lodocaine is used to stabilize the cardiac membrane in arrhythmias such as ventricular tachycardia by increasing the electrical stimulation threshold of the ventricles/His purkinje fibers.. This drug accomplishes this by precipitating a decrease in the hearts automaticity by blocking the sodium channels that feed the cells of the electrical conduction system of the heart. This occurs without precipitating profound changes in the different conduction intervals of the cardiac cycle. Lidocaine was developed in the year 1943 by a Swedish scientist named Nils Lofgren to be employed as a anaesthetic agent due to its nerve conduction blocking properties that produced a loss of feeling when used topically to block pain. This drug when employed in the cardiac setting is introduced into the system during cardiac arrhythmias through intravenous injection and is sometimes accompanied by epinepherine to aid in breaking the potentially fatal rhythm and to regain control of the heart's conduction system. Lidocaine needs to be monitored on telemetry for increased ventricular ectopy and any radical changes that may occur within the conduction cycle intervals.

Ventricular Septal Defect

Ventricular Septal Defect which also referred to as "VSD," is a hole or tear that is located within the septum between the right and left ventricle. This abnormal passageway can be the result of a couple of different factors. One possibility is that it is of a congenital nature that has been present from birth. This presents in up to 60% of infants born with heart defects of which approximately 85% of these defects close on their own shortly after birth. Another possible factor that could be considered is the possibility that it is a damaged and torn septum wall resulting from a heart attack (myocardial infarction). This condition compromises the heart's cardiovascular system in a couple of different ways that will in time present recognizable symptoms throughout the body and its organs. The increased amount of pressure generated within the left ventricle over the lesser pressure generated by the right ventricle during the depolarization phase of the cardiac cycle forces the blood from the left ventricle to shunt into the right ventricle. This mixes the already re-oxygenated blood with waste blood that is being sent to the lungs for re-oxygenation, presenting an excessive amount of blood presented to the lungs. The fluid overload increases the pressure within the right ventricular heart chamber, the pulmonary artery and the pulmonary vein, creating a condition called secondary pulmonary hypertension. This occurs when the lumens within the pulmonary vessels become smaller due to the thickening of their walls which results in diminished blood flow. The heart works harder to try and force an adequate amount of blood through the lungs for re-oxygenation to satisfy the body's needs and as a result can suffer heart failure, cardiomyopathy, or hypertrophy. This condition will also produce symptoms such as shortness of breath that can lead to episodes of pre-syncope and syncope brought on by inadequate oxygen supply to the brain. Severe instances of this Ventricular Septal Defect can eventually reduce the amount of physical activity that the individual will be able to engage in without experiencing fatigue or angina. The ventricular septal defect anomaly is diagnosed through the utilization of a few different methods. One such method would include the procedure called the cardiac catherization which utilizes a device that is inserted and fed through the femoral artery to the heart to measure the pressures within the right and left ventricles for a comparison. Other tests that might be helpful would be the ventricular angiogram which uses a contrast medium that is introduced into the ventricular chambers to present an image of their blood flow that is visible when used in conjunction with an x-ray, an echocardiogram which utilizes echo technology to present an real time image of the working heart on a viewing screen, or through cardiac auscultation where a stethoscope is used to hear audible sounds made by the blood as it travels through the VSD. Treatment for this condition depends on its severity and may consist of minimal treatment for smaller holes to surgical intervention in the form of stitching or patching to seal larger holes. However, it may be necessary to constitute lifestyle changes such as decreasing the physical activity level to accommodate the hearts ability to meet the body's demand and prevent further problems created by stress from overworking.

Fight or Flight

This is a natural physiological bodily response that is presented when there is a perceived possibility of injury in the form of physical or psychological damage. At times of duress the sympathetic nerves which are nerve fibers influence the SA node, AV node, atria and the ventricular muscle to become stimulated as a result of the release of norepinephrine from the medulla of the adrenal gland. This chemical produces an immediate elevation in the resting energy level by increasing the oxygen and blood sugar levels which produce enhansed myocardial contraction strength and endurance. This action increases the individual's heart rate and blood pressure which provides greater blood flow to the brain and muscles. For this fact, epinephrine is administered to individuals during episodes of cardiac arrest, asystole and PEA (pulseless electrical activity) to stimulate a myocardial response. This hormone was first isolated and named in the year 1901 by a U.S. chemist named Dr. Jokichi Takamine to be present in the secretions from animal adrenals. Epinepherine was first artificially reproduced in the year 1904 by the chemist Fredrich Stolz. Today this chemical is synthetically manufactured to produce the same physiological changes that would occur if it were released by the body's adrenal gland.

Did You Know?

Every year four hundred thousand individuals die from complications related to cigarette smoking. Smoking can triple your chances of becoming afflicted and dying from heart disease at mid-life. Smokers have a twenty-two to one higher probability of contracting and dying from lung cancer than a non-smoker, with the chance of contracting and dying from bronchitis and emphysema is increased by ten times.

Being overweight increases the risk of becoming afflicted with heart disease. It is estimated that sixty percent of the adults in the United States are obese and that forty percent engage in minimal or no physical activity. Studies have also shown that obesity in children has almost tripled in the past fifteen years.

"Tissue Plasminogen Activator" which is also referred to as TPA and clot buster, is a chemical that is composed of a protein similar to one that occurs naturally within the body. This substance is used to dissolve clots and is most effective within the first three hours of the onset of a heart attack or stroke. This frees the blockage and restores blood flow, but it does present an increased risk of bleeding.

Ventricular Tachycardia

Ventricular Tachycardia which is also referred to as "V-Tach" and "VT," is a fatal arrhythmia that occurs due to a conduction disturbance that occurs from within the ventricles. These episodes can be induced by scarring from dead or damaged cardiac tissue that is the result from a prior heart attack, enhanced cardiac cellular automaticity within the ventricles, electrical impulse circuit re-entry, or the result of triggered automaticity induced by a secondary source such as Long QT Syndrome. The ventricular tachycardia arrhythmia is classified into different groups that are catagorized by the evaluation of different factors such as its morphology, heart rate and time duration of the event. These classifications are the direct result of the process by which the ventricles conduct the electrical impulse. The appearance of the complex will determine if it is Monomorphic Ventricular Tachycardia or Polymorphic Ventricular Tachycardia. Monomorphic Ventricular Tachycardia will present with uniform complexes that possess the same appearance. This variation can be caused by factors such as increased automaticity that arises from a uni-focal point of origin or electrical impulses traveling within a re-entry circuit precipitated by damaged or dead non-conductive myocardial muscle. This conduction can cause an increased ventricular heart rate that overrides that of the normal conduction system. Polymorphic Ventricular Tachycardia which is also referred to as "Torsades de Pointes," is an arrhythmia that due to the varying amplitude displays different morphologies with each complex. This action presents its complexes to appear to be spiraling around the isoelectric baseline. This form of V-Tach is precipitated mainly by a congenital or drug induced anomaly referred to as Long QT Syndrome (LQTS), which is an extended repolarization phase of the ventricles. LQTS presents vulnerable cells within the relative or supernormal refractory period that can be prematurely stimulated by incoming electrical impulses that can induce the Torsades de Pointes arrhythmia through triggered automaticity. Ventricular Tachycardia generally possesses a heart rate between 101 and 250 beats per minute, however, when the heart rate is within 101-150 cycles per minute it is classified as slow V-Tach. Another factor that is used in the evaluation of this arrhythmia is the time duration of the event. When premature ventricular contractions (PVCs) occur in a run greater than 3 and produce a heart rate that exceeds 100 beats per minute but last for less than 30 seconds, is classified as a non-sustained run of V-Tach. However, if it continues for time period that lasts longer than 30 seconds it is called sustained V-Tach. Diagnosis of this arrhythmia is made mainly through the observation of the arrhythmia on cardiac telemetry or an electrocardiogram (EKG) tracing, but care must be taken when evaluating to be sure that there is no confusion with other abnormal conduction abnormalities such as Ashman's Phenomenon that during observation will seem to have the same appearance. It is best to assume that a Supraventricular Ventricular Tachycardia arrhythmia that can be possibly mistaken for a Ventricular Tachycardia arrhythmia be treated according to Ventricular Tachycardia protocol. This is due to the fact that certain medications such as calcium channel blockers that are used to treat SVT that are introduced into the system with Ventricular Tachycardia present can produce a negative effect. Otherwise, treatment would concentrate on the termination of the present life threatening arrhythmia. This is accomplished utilizing the following methods:

Defibrillation is the intentional introduction of an electrical shock through the chest wall to depolarize the myocardium. This momentary interruption within the heart's conduction system will terminate the arrhythmia and allow the sinoarial node to regain control and restore sinus rhythm when the heart restarts. After the arrhythmia is terminated there are a couple of different methods that are put in place to lessen the chances of the individual experiencing future recurrent episodes. Physicians may prescribe drug therapy such as beta-blockers to be taken by the individual in maintenance doses to prevent future arrhythmia episodes, or there is the option of the Automated Implantable Cardioverter Defibrillator (AICD) which is a device that is surgically placed within the chest of the individual who suffers from chronic repeat episodes. The AICD possesses a micro computer that continually senses the individual's rhythm and delivers an electrical shock to terminate the arrhythmia when it occurs.

Did You Know?

Epinephrine Adrenergic Cardiac Stimulant

This is a synthetically produced drug that mimics the hormone released by the medulla of the adrenal gland. This drug stimulates the heart giving increased muscular strength and endurance. Its use is indicated in situations where the use of electrical shock to stimulate myocardial response will have no effect such as asystole when the heart is in a depolarized state; or where there are electrical impulses present but no mechanical myocardial function (PEA). Adrenalin (Epinephrine) is a chemical that is secreted by the medulla of the adrenal gland. This hormone was first isolated and named in the year 1901 by a U.S. chemist named Dr. Jokichi Takamine to be present in the secretions from animal adrenals. Adrenalin is normally released in times of fear or injury to prepare the body for physical or mental stress producing what is referred to as the body's natural "Fight or Flight" response. Today, this chemical is synthetically manufactured to produce the same physiological changes. Epinepherine produces an immediate increase in the resting energy level (basal metabolism) and blood sugar level, producing enhanced myocardial contraction strength and endurance, which increases the individuals heart rate, blood pressure, and perfusion to the body. For this fact epinepherine is administered to individuals during episodes of cardiac arrest and PEA (pulseless electrical activity) to stimulate a myocardial response.

Atropine Antidysrhythmic

Atropine is used to increase the heart rate in cases that involve extreme bradycardic heart rates by denying vagal stimulation. This action also allows an increase the cardiac output. Atropine can increase ventricular ectopy and precipitate a tachycardic heart rate. Atropine is an alkaloid extract that comes from the plant that is referred to as the "Deadly Nightshade" or "Bella-Donna" which means beautiful lady. This phrase comes from to the idea that at one time in history men regarded women who used it to dilate their pupils for cosmetic reasons more attractive. This plant is poisonous enough to kill an adult with the ingestion of only one leaf and was used as a means of killing individuals by poisoning as recorded in history. From a cardiac standpoint, it is used to deny vagal stimulation to allow an increase in the heart rate in cases involving extreme bradycardia.

Wandering Atrial Pacemaker

Wandering Atrial Pacemaker which is also called WAP, is a supraventricular arrhythmia that is governed by the random firing of electrical impulses of varying amplitude. These impulses are delivered by multiple ectopic foci originating from different sites within the myocardium such as the sinoatrial node, the AV junction and the atria. As the vagal tone changes within the sinoatrial node (SA node) it slows its electrical impulse discharge to a level that the automaticity present in the atria and the atrioventricular node (AV node) surpass its discharge rate and they create and deliver their own electrical impulse. They discharge in an unregulated fashion with each different morphology will possess their own P-wave appearance, PR interval and R-R interval. This arrhythmia can be sustained and last for long periods of time, or it may be of a transient nature entering and leaving another base rhythm randomly. Wandering Atrial Pacemaker can be induced by respiratory factors such as those that are related to Chronic Obstructive Pulmonary Disease (COPD) which involves Chronic Bronchitis and Emphysema, or it may be induced by irritated or inflamed atrial tissue, electrolyte imbalance, or glycoside toxicity. Due to its variable R-R intervals, wandering atrial pacemaker appears similar to and can be confused with the cardiac arrhythmia atrial fibrillation. These two arrhythmias can possess the same characteristic irregularity, with the difference being that atrial fibrillation displays no discernible P-wave. This makes it necessary when evaluating these forms of arrhythmias to pay particular close attention for the presence of differing P-wave morphologies. Treatment for this arrhythmia would center on correcting the abnormality that precipitates the behavior.

The guidelines and characteristics that classify to arrhythmia are as follows:

* the arrhythmia will display irregularity.
* there will be one P-wave, per one QRS complex, per each cardiac cycle.
* there must be at least three different P-wave morphologies present on the strip being interrogated for this arrhythmia to be classified as wandering atrial pacemaker. The different morphologies will be unique to their specific site of the electrical impulse focal point of origin.
* the heart rate for wandering atrial pacemaker is under 100 beats per minute, with a heart rate greater than 100 beats per minute changing the classification of the arrhythmia to "Multi-Focal Atrial Tachycardia" or "MAT."
* the PR interval and R-R interval time durations will be variable, but will stay consistent for each individual morphologies point of origin.

Did You Know?

Being overweight increases the risk of becoming afflicted with heart disease. It is estimated that sixty percent of the adults in the United States are obese and that forty percent engage in minimal or no physical activity. Studies have also shown that obesity in children has almost tripled in the past fifteen years.

Wolff Parkinson White Syndrome

Wolff Parkinson White Syndrome which is also referred to as "WPW," is a condition that is induced by a physiological alteration within the myocardium that creates an accessory electrical impulse pathway adjacent to the site of the heart's natural AV nodal conduction pathway. This abnormal pathway is referred to as the "Kent Bundle" and is believed to be of a congenital nature, indicating that it has been present from birth. In the normal conduction system of the heart the electrical impulse leaves the sinoatrial node and travels along a prescribed route through an intra-nodal pathway to the atrioventricular node where it is regulated by the atrioventricular node as it is transmitted from the atria to the ventricles. This process produces a regular and controlled series of events that is referred to as the cardiac cycle, initiating timed muscular contractions that produce a heart rate between 60 and 100 beats per minute. The Kent Bundle presents itself as a direct link from the atria to the ventricles and does not possess an electrical impulse regulator (AV node). This absence of nodal control allows the unrestricted flow of erratic electrical impulses that have the capability of inducing heart rates that can reach as high as 250 beats per minute. This rapid heart rate is precipitated by a phenomenon referred to as "Pre-Excitation Syndrome." Pre-Excitation Syndrome occurs when the electrical impulse delay that normally occurs in the hearts conduction cycle which allows the atria to empty their fill of blood and the ventricles to refill does not happen. This unregulated delivery of electrical impulses to the ventricles causes them to depolarize prematurely and contract before they have received their complete fill of blood, diminishing the volume that is delivered to the body. Eventually this situation progresses to conditions such as ischemia/hypoxia, affecting the organs and tissue through oxygen deprivation that after time could result in necrosis. A few of the symptoms associated with this phenomenon could include lightheadedness, syncope, palpitations, shortness of breath and chest pain. Wolff Parkinson White Syndrome is recognizable on an electrocardiogram (EKG) through the observation of the presence of a Delta wave in lead V2. This wave is seen as an upstroke swell at the beginning of the QRS cycle that shortens the PR interval and creates a wide QRS complex.

In extreme situations involving sustained tachycardia arrhythmias, a procedure called Cardioversion is utilized to intentionally depolarize and reset the heart's conduction system. A timed electrical shock is introduced into the heart to stop all electrical activity in an attempt to allow the heart the opportunity to restart in normal sinus rhythm. Invasive procedures that would be employed to correct the problem would include open-heart surgery to repair the defect, or a minimally invasive procedure referred to as radiofrequency catheter ablation. This method of treatment utilizes radiofrequency energy to burn tissue and to close the accessory pathway. The ablation procedure is performed in an electrophysiology (EP) lab by an Electrophysiologist who specializes in the diagnosis and treatment of abnormalities of the conduction system of the heart. When successful, the ablation procedure results in a less than five percent rate for arrhythmia recurrence.

Frequently Used Cardiac Drugs and Their Application

Cardiac Drugs and Their Application

Adenosine (Adenocard) **Antidysrhythmic Endogenous Nucleoside**

Used to slow the conduction in tachycardic supraventricular arrhythmias by causing an interruption at the site of the atrioventricular node (induced 2nd degree AV block). This interruption provided the opportunity for the sinoatrial node to regain control of the conduction system within I minute in 60% of patients. Adenosine will produce P-wave asystole for a short time period, with the rhythm returning to normal sinus. The patient will need to be observed for EKG changes in the PR, QRS and QT intervals.

Amiodarone (Cordarone) **Antidysrhythmic Class lll**

This class lll medication extends the time duration of the refractory period, prolongs the PR and QRS intervals, and slows the sinus rate. Amiodarone is administered to aid in regaining control of supraventricular tachycardia arrhythmias, and is utilized in the chemical cardioversion of arrhythmias such as atrial fibrillation. Amiodarone was developed in the year1961 in Belgium by Dr. Bramah Singh to be used primarily as a treatment for angina. During it's use as an anti-angina medication, amiodarone was found by Dr Singh to be equally effective in the suppression of both ventricular and supraventricular arrhythmias due to its prolongation of the hearts repolarization cycle and QT interval, slowing of the electrical impulses as they travel through the heart, and it's effectiveness in decreasing the hearts automaticity or myocardial automatic memory reponse. It's use produced good success among his patients. With this news an Argentinean physician named Dr. Mauricio Rosenbaum utilized amiodarone on his patients, while keeping records of his successes. These notes eventually made their way to physicians in the U.S .in the 1970's, though due to non-acceptance by the FDA , their supplies had to be obtained in Canada. Approval by the FDA came about in the later part of 1985, classifying it as a Class lll Antidysrhythmic. Amiodarone has multiple side effects that in some instances can prove fatal, but due to its effectiveness and success in converting rhythm where other antidysrhythmics have failed to produce results, it is commonly used.

Atropine Antidysrhythmic

Atropine is used to increase the heart rate in extreme bradycardic situations by denying vagal stimulation reflexes in the myocardium. This action also increases the cardiac output. Atropine can increase ventricular ectopy and precipitate a tachycardic heart rate and patients on this drug need to be monitored closely. Atropine is an alkaloid extract that comes from the plant that is referred to as the "Deadly Nightshade" or "Bella-Donna" which means beautiful lady. This phrase comes from to the idea that at one time in history men regarded women who used it to dilate their pupils for cosmetic reasons more attractive. This plant is poisonous enough to kill an adult with the ingestion of only one leaf, and was used as a means of killing individuals by poisoning, as recorded in history. Atropine is used by optometrists to dilate pupils, which occurs due to its paralyzing effect on the iris and ciliary muscles. It is also used to limit the amount of oral and airway secretions produced before the administration of anesthesia. From a cardiac standpoint, it is used to deny vagal stimulation to increase the heart rate in cases involving extreme bradycardia.

Betapace (Sotalol) **Antidysrhythmic Class II**

This drug extends the time duration of the ventricular repolarization period and prolongs the QT interval. It is also used in treatment of ventricular dysrhythmias. Betapace needs to be monitored for EKG rhythm changes related to the ventricles such as Torsades de Pointes, the precipitation of heart blocks, or a bradycardic heart rate.

Corvert (Ibulitide) **Antidysrhythmic Class III**

Ibutilide is an intravenousis drug that is used to rapidly convert the arrhythmias atrial fibrillation/atrial flutter back to normal sinus rhythm in cases involving recent post CABG patients.; a practice that is done within a one week period of surgery with new onset atrial fibrillation / atrial flutter rhythm changes. This medication controls the hearts rhythm by regulating the synthesis of potassium and sodium with the hearts cells that are the base ingredients for the hearts electrical signals. By controlling this transition, the heart beat is kept regular. Patients on this medication need to be monitored for rhythm changes such as induced Torsades de Pointes that is precipitated by a lengthened QT interval, or the occurrence of rapid heart rates.

Coumadin (Warfarin) **Anticoagulant**

Coumadin is classified into a category of anticoagulants or "blood thinners," but is in actuality not a blood thinner. Coumadin is used as an anticoagulant in situations where there is the potential of clots forming present. This drug works as a clot prevention option that represses the synthesis of blood clotting factors that normally combine when there is a break in a blood vessel to stop the bleeding. Although coumadin is successful in helping to prevent clots it does not actually thin the blood, and cannot dissolve a clot that is already formed This drug is administered in situations involving arrhythmias such as atrial fibrillation where there is blood pooling due to a decreased circulation, prevention and management of DVT or deep vein thrombosis where there is the possibility of clots forming within the deep veins of the legs that can produce a pulmonary emboli. The major side effect of this medication is uncontrolled bleeding due to the induced inhibition of clotting factors.

Dobutamine (Dobutrex) **Adrenergic Cardiac Stimulant**

Dobutamine is administered in IV form to increase cardiac output without showing an increase in the heart rate through enhanced contractility. This drug is used to help compensate inadequate function of the heart by increasing the force of the hearts contractions to perfuse more blood volume in situations that involve valvular or myocardial defects, heart failure, or cardiac surgery. The individual is monitored for increased ectopic activity and tachycardic heart rates.

Digoxin (Digitalis) **Antidysrhythmic, Cardiac Glycosoide**

This drug is comprised of a chemical that is extracted from a plant called the "Digitalis Lanata." Digoxin enhances the power of the hearts contractions, providing for greater pushing force and increased blood perfusion to the body. Its properties also regulate the heart rhythm by decreasing the speed of the conduction at the site of the atrioventricular node. These properties were discovered in the year 1785 when William Withering, who was a British scientist, found it useful in treating certain diseases of the heart. Digoxin needs to be monitored for possible side effects related to dysrhythmias such as AV heart block, or drug induced bradycardia.

Dopamine Adrenergic **Cardiac Stimulant**

This drug is administered in IV form, and is used to increase cardiac output. It's effects on the heart can be seen in the EKG to possibly precipitate a widened QRS complex, increase ectopy, or to induce a tachycardic heart rate. The patient that is receiving this medication must be monitored for EKG changes.

Epinephrine Adrenergic **Cardiac Stimulant**

This is a synthetically produced drug that mimics the hormone released by the medulla of the adrenal gland. This drug stimulates the heart giving increased muscular strength and endurance. Its use is indicated in situations where the use of electrical shock to stimulate myocardial response will have no effect such as asystole when the heart is in a depolarized state; or where there are electrical impulses present but no mechanical myocardial function (PEA). Adrenalin (Epinephrine) is a chemical that is secreted by the medulla of the adrenal gland. This hormone was first isolated and named in the year 1901 by a U.S. chemist named Dr. Jokichi Takamine to be present in the secretions from animal adrenals. Adrenalin is normally released in times of fear or injury to prepare the body for physical or mental stress, producing what is referred to as the body's natural "Fight or Flight" response. Today, this chemical is synthetically manufactured to produce the same physiological changes. This drug produces an immediate increase in the resting energy level (basal metabolism) and blood sugar level, producing enhanced myocardial contraction strength and endurance, which increases the individuals heart rate, blood pressure, and perfusion to the body. For this fact it is administered to individuals during episodes of cardiac arrest, asystole and PEA (pulseless electrical activity) to stimulate a myocardial response.

Flecainide (Tambocor) **Antidysrhythmic** **Class 1C**

Utilized to stabilize cardiac membrane by decreasing the conduction in the purkinje fibers within the ventricles and throughout the hearts entire conduction system. The action produced by this drug has a positive effect on the management of ventricular tachycardia and supraventricular tachycardia arrhythmias when used. Patients need to be monitored for EKG changes in the PR, QRS and QT intervals of the cardiac cycle, the precipitation of heart blocks, increased ectopy or possible cardiac arrest.

Heparin Anticoagulant/ Antithrombotic

Heparin is a true anticoagulant because it becomes part of / combines with the blood to prevent or delay clotting. This substance is found in different body tissue, but mainly the liver, and is used as an agent to prevent coagulation and the formation of clotting in arrhythmias such as atrial fibrillation or conditions like deep vein thrombosis that can precipitate a pulmonary emboli. It is also used as a anti-coagulation agent in procedures such as open heart surgery to inhibit clotting. Although heparin will not dissolve a clot that is already formed, this medication is used as a theraputic measure in atrial fibrillation conditions such as chronic DVT to prevent clot formation due to the lack of blood volume movement.

Lidocaine (Parenteral) **Antidysrhythmic Class 1B**

Lodocaine is used to stabilize the cardiac membrane in arrhythmias such as ventricular tachycardia by increasing the electrical stimulation threshold of the ventricles/His purkinje fibers.. This drug accomplishes this by precipitating a decrease in the hearts automaticity by blocking the sodium channels that feed the cells of the electrical conduction system of the heart. This occurs without precipitating profound changes in the different conduction intervals of the cardiac cycle. Lidocaine was developed in the year 1943 by a Swedish scientist named Nils Lofgren to be employed as a anaesthetic agent due to its nerve conduction blocking properties that produced a loss of feeling when used topically and blockd pain when used in an epidural during events such as child birth. This drug when employed in the cardiac setting is introduced into the system during cardiac arrhythmias or cardiac arrest through intravenous injection that is sometimes accompanied by epinepherine to aid in breaking the potentially fatal rhythm to regain control of the hearts conduction system. Lidocaine needs to be monitored on telemetry for increased ventricular ectopy and any radical changes that may occur within the conduction cycle intervals.

Propranolol Antidysrhythmic Class ll

Used in the management of dysrhythmias by controlling the way the myocardium responds to stimuli. This medication is also used in the management of recurring chest pain. Patients that have been administered this drug need to be monitored on telemetry for occurrences such as heart block and drug induced bradycardia.

Quinidine Antidysrhythmic Class 1A

This drug is related to the substance called Quinine that is extracted from the bark of the chincona tree found in South America. Quinidine possesses the same chemical make-up, but also possesses properties that diminish the transmission of the nerve impulses, and lessen the excitability within the myocardium by extending the refractory period. This drug is used in the correction and managemant of arrhythmias such as atrial fibrillation and atrial flutter, paroxysmal atrial tachycardia, and ventricular tachycardia. This medication can induce bradycardia and needs to be monitored for EKG changes in the time duration periods of the different phases of the cardiac cycle for infusion rate adjustments.

Rythmol (Propafenone) **Antidysrhythmic Class 1C**

Reduces the reaction of the myocardial membrane, slows the conduction speed, and represses the hearts automaticity. This drug is utilized not only in the management of potentially fatal arrhythmias such as ventricular tachycardia, but is useful in preventing the return of rhythms such as atrial fibrillation and in blocking the conduction through accessory pathways in arrhythmias such as WPW (Wolf Parkinson White Syndrome). Patients that have been given this medication need to be monitored for EKG changes such as bradycardia, atrioventricular blocks, and abnormal conduction impulse delays within the ventricles.

Lopressor (Metoprolol) **Antianginal**

Lopressor is utilized mainly to lower blood pressure and heart rates, and to reduce cardiac related chest pain. This medication can precipitate cardiac arrest, bradycardic heart rates and heart blocks. Lopressor needs to be monitored on telemetry during and following IV delivery of medication for EKG changes.

Verapamil **Anti-Anginal** **Calcium Channel Blocker**

Decreases conduction at the SA/AV node. Used in correction and management of dysrhythmias and unstable angina. This drug when used requires monitoring for changes in the cardiac cycle intervals, bradycardic heart rate, and the precipitation of AV blocks.

Antidysrhythmic Drug Classification Chart

Class I	**Class IA**	**Class IB**	**Class IC**
Moricizine	Disopyramide	Lidocaine	Flecainide
	Procainamide	Mexiletine	Indecainide
	Quinidine	Phenytoin	Propafenone
		Tocainide	

Class II
Acebuterol
Esmolol
Propranolol
Sotalol

Class III
Amiodarone
Bretylium
Ibutilide

Class IV
Verapamil

Other medications utilized in cardiac treatment
Drug

Atropine

Notes

Terminology

Cardiac Terminology

Aberrancy a deviant myocardial response that produces no perfusion, different from that which is considered normal.

Ablate to remove by surgery, melt, vaporize, or wear away.

Ablation the process of destroying or repairing defective tissue through the use of a minimally invasive procedure such as a radiofrequency ablation which utilizes radiofrequency waves to burn away tissue.

Acetylcholine an alkaloid present within the bodily tissue that lowers the heart rate and blood pressure that becomes more pronounced when induced by signals from the vagus nerve. Related to the para-sympathetic nervous system.

Aneurysm an enlarged sac formed in the compromised wall of an artery by that has been weakened by disease or injury from a trauma.

Angina chest discomfort in the form of pain or squeezing pressure that can radiate to the shoulders and jaw that is commonly precipitated by ischemia that is induced by coronary artery disease (CAD).

Angiogram an minimally invasive exploratory procedure that utilizes the introduction of a contrast dye into the circulatory system via a catheter tube to evaluate the blood volume that is supplied to the heart through its arteries.

Angioplasty a minimally invasive procedure that is performed to re-establish blood flow to the heart muscle. This procedure utilizes an inflatable balloon to open blood vessels that have been narrowed due to the build-up of plaque and fatty deposits.

Anti-Coagulants chemicals designed to inhibit thrombin and blood clotting factors that are introduced into the circulatory system to aid in the suppression of the formation of clots.

Anticoagulate the act of introducing anti-coagulating medication into the circulatory system to inhibit clotting factors within the blood.

Aorta the main artery that carries blood away from the heart to be delivered to smaller arteries to feed the body. This artery receives its blood from the left ventricle.

Aortic Valve a semi-lunar valve located between the left ventricle and the aorta.

Arrest complete failure of the sinoatrial node to generate an electrical impulse for a period of three seconds or greater. This event leaves the myocardium in a depolarized state.

Arrhythmia abnormal heart rhythm that is induced by defects within the conduction system that causes it to be atypical to that which is considered to be normal sinus rhythm.

Arterial Blood Gases (ABG's) a test that measures various levels such as Ph, O2 and CO2, that are present within freshly re-oxygenated blood drawn from an artery.

Artery thick walled vessels that carry blood away from the heart to be delivered to the body's organs and lungs.

Asystole the complete cessation of all electrical and muscular cardiac activity.

Atherosclerosis a condition in which the arterial pathways are narrowed due to the accumulation of plaque and fatty deposits that results in a restriction of blood flow. This disease leads to ischemia which will eventually cause damage or death of the body's cells and tissue.

Atria the two upper chambers of the heart muscle that are referred to as the right atria and the left atria.

Automaticity the ability of cardiac cells to create their own electric impulses to depolarize the myocardium without outside intervention.

AV Block (Atrioventricular Block) the delay or complete block of the electrical impulse as it travels through the AV node from the atria to the ventricles.

AV Junction the section of the heart where the electric impulse is transmitted from the atria to the ventricles that is located between the upper and lower chambers of the heart. This section contains the AV node and Bundle of HIS.

AV Node a cell cluster located in the conduction pathway of the heart that regulates and transmits the electrical impulse as it travels from the atria to the ventricles.

Beta Blockers medications formulated to reduce blood pressure to lessen the workload on the heart muscle, commonly used in situations involving congestive heart failure and angina.

Bi-Ventricular used in pacemaker terminology to describe the synchronization of the electrical impulse delivered to the two lower chambers of the heart to ensure simultaneous depolarization.

Bradycardia classified as a heart rate of less than sixty beats per minute.

Bundle Branches bundles of cells that transmit electrical impulses from the atria to the ventricles that are located below the Bundle of His in the pathway of the hearts conduction system. This bundle divides into two branches, one into the right, and one into the left ventricle.

Bundle Branch Block (BBB) an abnormality present within either the right or left bundle branch of the conduction system that delays the transmission of the electrical impulse as it attempts to pass through from the atria to the ventricles.

Cardiac Cycle one complete cycle of the heart, from the initial impulse sent by the sinoatrial node to the completion of the repolarization of the ventricles.

Cardiology the sector of medicine that pertains to function of the heart muscle, with relevance to the diagnosis and treatment of genetic or acquired conditions and diseases.

Cardiomyopathy terminology used to describe the loss of flexibility within the tissue of the myocardium due to the stiffening, thickening, or enlargement caused by factors such as uncontrolled hypertension that overworks the heart muscle.

Cardioversion a procedure employed to convert arrhythmias such as atrial fibrillation back to sinus rhythm by utilizing a controlled electrical shock to depolarize the myocardium. This creates a momentary window to provide an opportunity for the sinoatrial node to re-establish itself as the hearts pacemaker.

Catherization a minimally invasive exploratory procedure performed to evaluate the condition of the coronary arteries and blood flow to the heart muscle.

Catheter a thin tube that is inserted into the body to promote drainage, or allow the delivery of contrast dye and other various implements utilized during medical procedures such as ablation and angioplasty.

Chronic constant, recurring, lasting a long time.

COPD stands for Chronic Obstructive Pulmonary Disease which consists of two different lung diseases that cause difficulty breathing and diminish oxygen levels within the body by obstructing the airflow. It may consist of emphysema or chronic bronchitis individually, or together.

Conduction the transmission of an electrical impulse from one location to another.

Congenital present from birth, inherent.

Contraction the squeezing of the hearts atrial or ventricular chambers as a reflex action to the stimulation of an electrical impulse.

Depolarization the muscular response to electrical stimulation in the form of contractions, in which the blood is forced from the atria to the ventricles, or from the ventricles to the lungs and body.

Diastolic the resting phase of the chambers of the heart that occurs after the depolarization phase of the cardiac cycle.

Dyspnea shortness of breath with painful breathing.

Echocardiogram a non-invasive test that utilizes echo technology through the use of high frequency sound waves to create a visible image of the heart when evaluating for improper function and blood flow.

Ectopy abnormal cardiac electrical conduction activity, commonly seen as premature ventricular contractions (PVC's) or premature atrial contractions (PAC's)

Effusion the unintentional introduction of fluid into a body cavity or tissue.

Ejection Fraction the amount of blood forced from the heart's left ventricle during the depolarization phase of the cardiac cycle. This is measured in the volume delivered as compared to the remaining volume within the hearts chamber.

Electrocardiogram (EKG/ECG) a non-invasive test that utilizes twelve electrodes placed on various locations of the body to record the electrical impulse as it travels through the conduction system of the heart.

Electrophysiology the study related to the diagnosis and treatment of abnormalities present within the conduction system of the heart.

Embolism a blood clot, globule of fat, air bubble, piece of plaque, bone marrow or tissue traveling through the blood stream.

Endarterectomy a surgical procedure in which the carotid arteries within the neck are opened and cleaned of plaque accumulation to re-establish blood flow to the brain.

Enzymes proteins present in all cells that initiate and facilitate the body's chemical reactions.

Epinepherine a hormone that is secreted by the medulla of the adrenal gland that enhances the strength and endurance of the muscles by increasing the oxygen and glucose levels within the blood. Related to the sympathetic nervous system.

Fusion Beat an event that occurs when the hearts intrinsic heart beat and the electric impulse sent to the myocardium by a pacemaker happen simultaneously.

Heart Failure a condition in which disease has affected the heart muscle to a degree that it is too weak to adequately provide catalyst to sufficiently move blood to the body and its organs.

His Bundle a specialized cluster of cells located at the AV junction that follows the atrioventricular node and precedes the right and left bundle branches.

Hypercalcemia elevated levels of calcium within the body.

Hyperkalemia elevated levels of potassium within the body.

Hypertension a condition that is created when there is consistent blood pressure greater than the normal high limits of 140/90.

Hypertrophy enlargement or thickening of the tissue of the heart muscle.

Hypocalcemia abnormally low levels of calcium within the body.

Hypokalemia abnormally low levels of potassium within the body.

Hypothermia a condition that presents when the body's core temperature drops below 95 F.

Hypoxia a condition that affects the body's organs/heart that is created due to a lack of oxygen.

Hysteresis a pacemaker pulse interval that is programmed to allow the heart every opportunity to produce its own intrinsic electrical impulse to depolarize the myocardium before it delivers an electric impulse.

Idioventricular ventricular only, no presence of electrical activity present above the ventricles.

Intrinsic generated by self, inherent, not dependant upon outside factors to function.

Invasive the entering or invasion into the body.

Ischemia a condition that is induced by the compromised blood flow, precipitating oxygen deprivation that can result in possible irreversible damage or death to the affected organs.

Isoelectric Baseline the line made on an EKG tracing during the inactive electrical periods of the cardiac cycle.

Kent Bundle the terminology used to describe an uncontrolled and unregulated conduction pathway that runs adjacent to the AV nodal pathway that is present in the arrhythmia known as Wolff Parkinson White Syndrome or WPW.

Lesion an injury or change in the bodily tissue or organs that commonly results in the reduced or complete inability of it to function properly.

Lumen the passageway located within a blood vessel.

Mitral Valve a valve with two cusps (flaps) that is located between the left atria and the left ventricle.

Multi-Focal describes electrical impulses that originate from different focal points within the myocardium.

Myocardial Infarction also called a heart attack, this is an event that occurs when the heart muscle experiences a diminished supply or stoppage of oxygen-rich blood flow which causes necrosis or death to the myocardial tissue.

Myocardium the part of the heart that makes up the muscular substance.
Myocarditis the inflammation of the myocardium.
Node a concentration of specialized cells that conduct and relay electrical impulses.
P-wave the wave form made by the depolarization of the atria on an EKG ttracing.
Palpitation heart related activity that produces the noticeable feeling of rapid beating, fluttering, galloping, quivering, or trembling.
Paroxysmal the inserting of one event into another. Presenting itself in short bursts such as an episode of Paroxysmal SupraVentricular Tachycardia (PSVT) which is an intermittent episode of an atrial induced heart rate that exceeds 160 beats per minute, happening in a short burst that can be seen to start and stop. Not sustained.
Perfusion the delivery of blood from the left ventricle through the bodily tissue and its organs.
Pericardial Window a surgical procedure that creates an opening in the pericardium to relieve pressure by allowing effusive fluid that has accumulated around the heart to drain.
Pericardialcentesis the intentional puncture of the pericardium to remove accumulated fluid.
Pericardium a thin fluid filled sac made of tough tissue that surrounds the heart that acts as a buffer to protect it from rubbing against the lungs and chest wall.
Pre-Syncope the feeling of dizziness and impending loss of consciousness.
Prolapse out of place, improper fitting, mitral valve prolapse.
Pulmonary Valve the two flapped semi-lunar valve that is located between the right ventricle and the right and left pulmonary arteries.
QRS complex three waves that form a pattern on the EKG tracing during the ventricular depolarization.
Regurgitation the leaking of blood in a reverse direction through a defective valve.
Repolarize the phase of the cardiac cycle following depolarization, when the heart rests and receives another fill of blood.
Restenosis the re-narrowing of a blood vessel following a PTCA procedure that was performed to open it and re-establish blood flow.
Retrograde reverse, moving in a direction that is contrary to the normal. A retrograde electrical impulse traveling in the reverse direction from the bundle of his back to the atria.
Septum partition, the thick layer of tissue that separates the right and left sides of the heart.
Sinus pertaining to the sinus mechanism of the conduction system of the heart.
Stent a device made of medicated wire mesh in the shape of a tube that is deployed with the use of a catheter to compromised locations in an artery to provide structural integrity and to ensure a vessel remains open for the passage of blood.
Sternum breastbone located in the chest over the heart.
Stroke a sudden neurological event precipitated by oxygen-rich blood deprivation or a cerebral hemorrhage event that has the ability to temporarily or permanently paralyze the body, affect vision and speech, create reflex disorders, or cause death. This event is commonly induced by a blockage in a cerebral artery.

Subarachnoid the space between the arachnoid membrane and the pia matter that surrounds the brain.

Subarachnoid Hemorrhage bleeding into the subarachnoid area that can induce a stroke.

Supra located above, supra-ventricular is a location above the ventricles.

Symptom a physical indication or response to a disease, metabolic imbalance, or condition present within the body that can be useful in pointing to the direction of potential problems when forming a diagnosis.

Syncope fainting, loss of consciousness caused by a temporary interruption in the blood supply to the brain.

Systolic the depolarization phase of the cardiac cycle when the hearts chambers are electrically stimulated to contract and squeeze to move blood from atrial chambers to the ventricular chamber and from the ventricular chambers to the lungs and body.

T-wave the wave pattern made on an EKG tracing during the repolarization of the ventricles

Tachycardia defined as a heart rate exceeding 100 beats per minute, that originating from the sinus node (sinus tachycardia) is generally between 100 and 160 beats per minute.

Tamponade increased pressure on the surface of the heart which compresses the surface arteries that feed the myocardium and impedes its chambers from expanding and filling with blood. This condition is also referred to as "Cardiac Tamponade" or "Pericardial Tamponade." This condition is caused by fluid build-up within the pericardium.

Telemetry equipment designed to receive, transmit, display and record electrical activity occurring within the heart. Information is delivered by a sending unit (transmitter) being worn by a patient to a receiving unit where it is converted to produce a wave form on a cardiac monitor at a central receiving station.

Thoracic pertaining to the torso or chest area of the body.

Thromboembolism an embolism that is carried by the blood through a blood stream that becomes lodged within a vessel.

Thrombolytics drugs used to dissolve and retard the formation of clots. Also referred to as "clot busters."

Thrombus blood clot comprised of fibrin and platelets that accumulates at the site of thrombosis that causes a blockage at its point of origin.
* Occluding Thrombus - blocks the entire lumen of a blood vessel.
* Parietal Thrombus – thrombus attached to a vessel or the heart wall.
* Mural Thrombus – thrombus attached to the wall of the endocardium.

Thrombylitics chemicals designed to dissolve clots that block arterial pathways that are responsible for inducing strokes and heart attacks. Also referred to as "Clot Busters."

Transient not permanent, intermittent, staying for a short time only. Transient second degree AV heart block is an arrhythmia that impedes or stops the electrical impulse communication at the site of the AV junction between the atria and ventricles in an intermittent fashion.

Tricuspid Valve a three flapped valve that is located between the right atria and the right ventricle.

TIA (Transient Ischemic Attack) short neurological events generally lasting for a time period of 8-14 minutes that are induced by temporary blockages in a cerebral artery.

Uni-Focal describes electrical impulses that present from one origin outside the hearts normal conduction pathway.

Vein a vessel that carries blood that has been stripped of its nutrients back to the right atrium of the heart to be sent via the right ventricle to the lungs for re-oxygenation.

Ventricles the two lower chambers of the heart that are referred to as the right ventricle and the left ventricle.

Pacemaker Terminology

AV Delay the time delay between the electrical impulse sent to depolarize the atria and the electrical impulse sent to depolarize the ventricle.

VA Delay the time duration between the ventricular depolarization in the previous cycle and the atrial depolarization in the new cycle.

Pulse Interval the total of the AV and the VA intervals in dual chamber pacemakers as measured in milliseconds.

Capture confirmation that an electrical impulse sent to the atria or ventricles has stimulated a depolarization within the proper location and timing.

Crosstalking this occurs when the impulse sent to the ventricles from a dual chamber pacemaker is retrograde conducted to the atria inducing a tachacardic heart rate.

Dual Chamber Pacing the sending of an electrical impulse to both the atria and the ventricles.

Electromagnetic Interference (EMI) outside interference that is caused by electrical or magnetic equipment that interrupts the normal operation of the pacemaker.

Electrical Output the amount of electrical current required to initiate a depolarization.

Fusion Beat this happens when the impulse sent by a pacemaker and the hearts own (intrinsic) beat happen simultaneously.

Hysteresis a prolonged pulse interval to allow the heart to provide it's own stimulation.

ICD (Implantable Cardioverter Defibrillator) this is a pacing device that is implanted in the chest that is designed to control rapid heart rates and to terminate Ventricular Fibrillation, Ventricular Tachycardia and Torsade de Pointes utilizing electrical shock.

Pulse Generator the part of a pacemaker that is designed to generate the electrical impulse that is used to deliver a shock to the heart muscle.

Rate Modulated pacemaker that senses the activity level of an individual and makes the heart rate adjustments automatically.

Single Chamber Pacing the sending of an electrical impulse to either the atria or the ventricle.

Standby Rate the lowest heart rate that a pacemaker will allow the intrinsic heart rate to go before sending impulses to initiate pacing.

Notes

Question & Answer Section

?

A series of questions with multiple choice answers to help enhance your learning experience, complete with a fully explanatory answer key at the end.

1. The heart has four chambers that are referred to as
   a. the upper chambers are the ventricles and the lower are called the atria
   b. the upper chambers are the atria and the lower are called the ventricles

2. The right and left chambers (right and left sides of the heart) are divided by a thick muscle that is referred to as the
   a. bundle of his
   b. septum
   c. bundle branch
   d. myocardial wall

3. The hearts intrinsic electrical impulse originates from specialized cells located in the upper section right atria that is referred to as the
   a. sinoatrial node
   b. atrioventricular node
   c. left atria

4. There are four valves that control the flow of blood through the heart as it passes from chamber to chamber, and as it exits the ventricles to feed the body and lungs
   a. true
   b. false

5. The valve that controls the flow of blood from the right atria to the right ventricle is referred to as the
   a. mitral valve
   b. tricuspid valve
   c. semi-lunar valve

6. The heart valves that control the flow of blood as it exits from the right and left ventricles as it is sent to the lungs and body are referred to as
   a. semi-lunar valves
   b. ventricular exit valves
   c. tricuspid valves
   d. check valves

7. The blood that is received into the right atria that is passed to the right ventricle is
   a. blood that contains carbon dioxide that is a waste product from the body that is sent through the pulmonary artery to the lungs where it is to be re-oxygenated
   b. re-oxygenated blood that is sent to the body

8. The correct normal intrinsic heart rate is
   a. 70-100 beats per minute
   b. 50-110 beats per minute
   c. 60-100 beats per minute

9. One entire process of the heart from the initial impulse originated at the site of the sinoatrial node to its completion in the purkinje fibers located in the ventricles (PQRST) is called
   a. sinus rhythm cycle
   b. cardiac cycle
   c. heart beat

10. The resting phase of the cardiac cycle that occurs during repolarization when the chambers are refilling with blood is referred to as diastole
    a. true
    b. false

11. The correct prescribed measurements for the normal sinus rhythm are
    a. PRI .16 - .22 QRS .04 - .10 QT .30 - .40
    b. PRI .14 - .20 QRS .06 - .12 QT .35 - .42
    c. PRI .12 - .20 QRS .06 - .10 QT .36 - .44

12. The correct inherent heart rates for the SA node, AV node / Bundle of His, and the Purkinje Fibers are
    a. 60 -100 / 40 -60 / 20 -40
    b. 60 -110 / 40 -60 / 25 -50
    c. 50 -100 / 45 -70 / 30 -50
    d. 75-125 / 50 - 74 / 25 -45

13. The R-wave to R-wave measurement located on a electrocardiogram tracing of a rhythm strip references to the time duration between the
    a. atrial depolarization cycles
    b. ventricular depolarization cycles
    c. atrial repolarization cycles
    d. ventricular repolarization cycles

14. A measurement of a PRI with a time duration that is greater than 0.20 indicates that there is a delay in the impulse as it travels through the conduction pathway at the point where it passes from the
    a. atria to the ventricle
    b. left atria to right atria
    c. left ventricle to right ventricle
    d. AV node to the SA node

15. A measurement of the time duration of a ventricular depolarization or QRS that is 0.14 sec. would indicate that there is a
    a. first degree AV block
    b. bundle branch block
    c. wenckebach
    d. premature atrial complex

16. There are two branches located in the conduction pathway that separate and extend to each ventricle immediately following the Bundle of His, they are referred to as the right and left bundle branches
    a. true
    b. false

17. The Electrocardiogram (EKG) is a test that is performed to
    a. measure the amount of blood flow that travels through the hearts chambers
    b. provide a tracing of the hearts electrical activity as the impulse travels from the sinoatrial node to the purkinje fibers
    c. provide the Physician with a view of the function of the hearts valves and chambers

18. The T-wave on a cardiac rhythm strip represents
    a. atrial depolarization
    b. atrial repolarization
    c. ventricular depolarization
    d. ventricular repolarization

19. Atrial Fibrillation is recognizable on a heart monitor by observing the following characteristic
    a. more than one P-wave per QRS per cycle
    b. irregularity with varying PRI measurements
    c. irregular heart rhythm with no discernable P-wave

20. Electrophysiology is the study related to the heart that is performed by an electrophysiologist to
    a. examine and treat abnormalities associated with the hearts valves
    b. diagnose and treat abnormalities that affect and/or are associated with the hearts conduction system
    c. the test performed to provide the Physician with a measurement in volume of the blood perfusion to the body during depolarization of the left ventricle

21. Sinus pause is a conduction abnormality that
    a. occurs in the rhythms of atrial fibrillation and atrial flutter that lasts more than three seconds
    b. an abnormality related to the sinoatrial node that momentarily prevents the formation of the electrical impulse
    c. failure of a ventricular depolarization to occur caused by diseased tissue in the atrioventricular node

22. Premature Ventricular Complexes are ectopic beats that are formed low in the ventricles that occur earlier than the normal beat that can happen only in normal sinus rhythm
    a. true
    b. false

23. The Junctional arrhythmia is identified on a cardiac monitor by observing that there is the presence of
    a. an upright and peaked P-wave that measures under a time duration of .12 seconds with a heart rate between 40 – 60 beats per minute
    b. an inverted P-wave that measures over .12 seconds and is located after the QRS, with a heart rate between 20 – 60 beats per minute
    c. an inverted P-wave that normally measures under .12 seconds that can fall before, within, or after the QRS, with a heart rate between 40 – 60 beats per minute
    d. a pacemaker spike that is not followed by a QRS

24. Ectopy is the term used to describe abnormal beats that occur in a rhythm that can include but are not limited to a) premature ventricular complexes (PVCs) b) premature atrial complexes (PACs) c) premature junctional beats (PJC's) d) sinus pause
    a. a, c, & d
    b. c & d
    c. a b & c
    d. all of the above

25. Arrhythmia is a term that is used to identify rhythms that display abnormalities that are atypical to the set guidelines used to determine normal sinus rhythm
    a. true
    b. false

26. Paroxysmal Supraventricular Tachycardia (PSVT) is used to describe
    a. arrhythmias that are induced by abnormalities that occur above the ventricles that you can see start and stop / happening in short runs, with a heart rate generally between 160 –250 beats per minute
    b. runs or group beating of multiple premature ventricular complexes consisting of more than five in a row
    c. ventricular tachacardia and ventricular fibrillation with a heart rate below 100 beats per minute

27. The electrical impulse as it travels through the heart is measured in
    a. centimeters
    b. time duration
    c. millimeter segments
    d. amplitude

28. The section on a rhythm strip that indicates that the ventricular depolarization has ended and repolarization has begun is the
    a. QRS
    b. T-wave
    c. J-point

29. The PR interval is representative of the time that it takes the impulse to stimulate a ventricular depolarization from the point where it initiated the atrial depolarization
    a. true
    b. false

30. A arrhythmia that displays a gradual lengthening of the PRI until there is a dropped QRS on the rhythm strip is referred to as
    a. second degree AV block type 1 wenckebach / mobitz 1
    b. first degree AV block
    c. second degree AV block type ll
    d. AV disassociation

31. Cardiac catherization is a minimally invasive information gathering procedure that includes the following test(s) (may be more than one answer)
    a. measure the blood pressures in the hearts major arteries
    b. taking of blood test samples
    c. conduct a coronary angiogram / inject a contrast medium to display abnormalities in the arteries on an x-ray screen
    d. do a left ventriculogram using a dye to provide information regarding the blood flow within the left ventricle
    e. all of the above

32. Cardiac enzymes are proteins that are present in cells that initiate and speed up the body's chemical reactions. Those that are specifically related to the heart muscle are referred to as CPK, CKMB, and Troponin. The purpose in studying these through a blood work-up is to evaluate whether a heart attack has occurred, and how severe it was as indicated through elevated test result levels
    a. true
    b. false

33. Cardiac arrest is an event in which
    a. there is no formation of an electrical impulse in the atrioventricular node for a period of four seconds or more
    b. there is no formation of an electrical impulse in the sinoatrial node for a period of three seconds or more
    c. there is no ventricular response to the electrical impulse stimulation for a period of three seconds or more
    d. there is no formation of an impulse in the sinoatrial node for a period of four seconds

34. Second degree AV block type ll displays which characteristics on a cardiac monitor
    a. constant PRI, more than one P-wave per QRS, regular atrial rhythm, ventricular rhythm can be irregular with bradycardic heart rate
    b. variable PRI, more than one P-wave per QRS, irregular atrial rhythm, regular ventricular rhythm with heart rate between 60 and 100 beats per minute
    c. irregular rhythm with frequent episodes of missed P-waves, bradycardic heart rate

35. In Third Degree AV Block, also referred to as "Complete Heart Block," there is a complete block of the conduction or transmission of the electrical impulse between the atria and the ventricle. This abnormality is precipitated as the result of
    a. diseased tissue in the ventricles cause them to not receive the impulse
    b. abnormalities in the atrioventricular node do not allow impulses through to the ventricles
    c. the septum becomes too thick for the impulses to pass through

36. In the arrhythmia Third Degree AV Block the ventricles possess an intrinsic protective mechanism within the Purkinje Fibers that allows them to function independently, the intrinsic heart rate for this occurrence would be
    a. 20 – 40 beats per minute
    b. 40 – 60 beats per minute
    c. 30 – 50 beats per minute
    d. 50 – 100 beats per minute

37. Sick sinus syndrome, also referred to as "Tacky Brady Syndrome," receives its name due to abnormalities in the SA node that cause sudden and abrupt changes in the heart rate from tachycardic to bradycardic (60 or below – 100 or above)
    a. true
    b. false

38. Torsades de Pointes, also called Polymorphic Ventricular Tachycardia, is a variant of Ventricular Tachacardia that is distinguishable on a cardiac monitor by recognizing the following observations (may be more than one answer)
    a. heart rate between 100 -125 beats per minute
    b. QRS morphology changes due to altering amplitude
    c. appearance of rhythm tracing spiraling around the isoelectric baseline
    d. more than one P-wave per QRS

39. Select the correct rhythm that possesses the following characteristics a) different appearances in the P-waves b) different PRI measurements c) different R-R intervals d) a heart rate greater than 100 beats per minute
    a. Wandering Atrial Pacemaker (WAP)
    b. Second Degree AV Heart Block Type 1
    c. Multi-Focal Atrial Tachycardia (MAT)
    d. Uncontrolled Atrial Fibrillation

40. Ventricular fibrillation is a fatal arrhythmia governed by abnormal and disorganized electrical activity that produces no perfusion to the body. This arrhythmia requires an electric shock to create an interruption in this activity to regain control of the hearts conduction system and return it back to the control of the sinoatrial node
    a. true
    b. false

41. Ejection fraction is terminology that describes
    a. the amount of blood perfused from the right ventricle during depolarization to the lungs for re-oxygenation
    b. the amount of blood that is perfused to the body during the depolarization of the left ventricle measured by volume
    c. the amount of blood passed from the atria to the ventricle in one heart cycle
    d. all of the above

42. The condition created when the body does not receive enough oxygen-rich blood to meet its needs is referred to as
    a. hysteresis
    b. hypoxia
    c. hyperkalemia
    d. dyspnea

43. A blocked Premature Atrial Contraction (PAC) is an ectopic atrial beat that is recognizable on a cardiac monitor by observing that
    a. the atrial depolarization came early
    b. the PRI's before the event were uniform to each other
    c. there was an atrial depolarization without a ventricular depolarization following it
    d. both a & c
    e. all of the above

44. Syncope is the body's symptomatic response in the form of fainting/loss of consciousness caused by a temporary deficiency of oxygen-rich blood supply to the brain
    a. true
    b. false

45. An electrical impulse created within the ventricles (purkinje fibers) in the absence of a conducted electrical impulse from the atria is referred to as
    a. wenckebach
    b. first degree atrioventricular heart block
    c. ventricular escape beat
    d. intrinsic ventricular activity

46. Ischemia is the damage to the hearts tissue that is caused by a lack of oxygen-rich blood
    a. true
    b. false

47. Pacemaker failure to sense is a pacemaker malfunction that occurs when
    a. there is an electrical impulse sent by the pacemaker, but no heart response following it
    b. the impulse sent by the pacemaker is timed in the wrong location of the cardiac cycle
    c. there is no impulse sent due to a malfunctioning pulse generator
    d. the impulse sent to the atria caused a ventricular response

48. In using pacemaker terminology, match the following terms with the proper definition
1) AV delay 2) VA delay 3) Pulse Interval 4) Fusion Beat 5) Failure to Capture
__ represents the time duration from the atrial impulse to the ventricular impulse
__ the impulse sent by the pulse generator failed to initiate any cardiac response
__ represents the time duration from the ventricular impulse in the present beat to the atrial impulse in the following beat
__ this is the total time duration as measured in milliseconds of the AV & VA delay
__ in this situation the pacing impulse and the hearts intrinsic beat occurred at the same moment

49. Select all that apply to Wolff Parkinson White Syndrome (WPW)
a. is also called pre-excitation syndrome
b. is caused by an abnormal adjacent pathway utilized by the electrical impulse other than the hearts normal impulse pathway
c. can cause rapid heart rates as a result of a reentry stimulation
d. produces an irregular rhythm due to unregulated impulses sent to the ventricles
e. keeps the rhythm regular with a bradycardic heart rate possessing more P-waves than QRS's
f. this abnormality is believed to be congenital (present since birth)
g. answers b, c, & d
h. answers a, b, c, d, f
i. all of the above

50. This occurrence happens when the impulse from a dual chamber pacemaker is sent to the ventricles and is retrograde conducted back to the atria induing a tachycardic heart rate (pacemaker mediated tachycardia)
a. hysteresis
b. cross impulse
c. crosstalking
d. dual chamber fusion

51. **part 1** This test utilizes echo technology with high frequency sound waves to produce the hearts image on a viewing screen
a) echocardiogram b) catherization c) x-ray d) EKG

part 2 This same test is performed to provide the Physician the following information
(check all that do not apply)
a. information on the hearts size
b. information on the hearts location
c. information on the amplitude of the hearts impulse
d. chamber and valve function
e. direction and velocity of blood flow
f. congenital abnormalities
g. all of the above

52. The sustained heart rate for ventricular tachycardia is 150 – 250 beats per minute, what is the heart rate considered to be " slow ventricular tachycardia"
    a. 75 – 100 beats per minute
    b. above 100 but less than 150 beats per minute
    c. 50 - 100 beats per minute
    d. all of the above

53. Long QT Syndrome, also called Prolonged QT Syndrome, is the phrase used to describe a heart condition that possesses which of the following characteristics (may be more than one answer/circle all that apply)
    a. more than one P-wave per QRS
    b. prolonged ventricular repolarization cycle (QT-interval)
    c. failure of the QT segment in the cardiac cycle to shorten with an increased heart rate
    d. bundle branch block that measures over 0.20 second

54. When manually calculating the intervals on a telemetry strip, the prescribed time durations of the boxes imprinted on the strip would be
    a. small boxes 0.05 large boxes 0.25
    b. small boxes 0.10 large boxes 0.50
    c. small boxes 0.04 large boxes 0.20

55. This procedure is performed in situations where there are arrhythmias that produce dangerously high heart rates (ex. rapid atrial fibrillation) that require intervention in the form of an electric shock to create an interruption and return the conduction of the heart back to its natural pacemaker
    a. transient electrical stimulation
    b. electrical cardioversion
    c. radiofrequency catheter ablation
    d. electrocardiogram

56. Is a three second and longer time duration between cardiac cycles considered a sinus arrest in normal sinus rhythm, atrial fibrillation, and atrial flutter
    a. yes / all rhythms
    b. no / sinus rhythm only
    c. atrial fibrillation / atrial flutter only
    d. atrial flutter only

57. Anti-coagulation therapy is medication used to thin blood and inhibit clotting factors to prevent coagulation (clotting of blood) that pools in the heart due to lack of movement. There are specific drugs that are utilized by Physicians to produce anti-coagulation properties within the blood, they are
    a. diovan htc
    b. coumadin
    c. amiodarone
    d. heparin

58. This treatment is designed to treat patients with heart failure in situations when conduction abnormalities cause the ventricles to lose their synchronism / or ability to contract together. This process / procedure stimulates both ventricles to depolarize at the same time (may be more than one answer / select all that apply)
    a. implementation of a bi-ventricular pacing
    b. is referred to as cardiac resynchronization therapy
    c. implementation of a dual chamber pacemaker
    d. implementation of a implantable cardioverter defibrillator
    e. all of the above

59. Match the correct definition with the proper terminology

_ embolus
_ intrinsic
_ repolarize
_ paroxysmal
_ ischemia
_ hyperkalemia
_ aneurysm
_ congenital
_ hysteresis
_ fusion beat
_ PSVT
_ bradycardia
_ diastolic
_ dyspnea
_ retrograde
_ hypertrophy
_ standby rate
_ transient
_ rate modulated
_ ablate
_ conduction
_ invasive
_ thrombus
_ cardiomyopathy
_ angina
_ hypertension

a. heart rate below the normal guidelines of 60 – 100 bpm
b. enlargement of the heart muscle
c. arrhythmia that temporarily enters and leaves a rhythm
d. shortness of breath
e. resting phase of the chambers of the heart
f. globule of fat or born marrow, piece of tissue, broken off plaque, or air bubble traveling through the blood stream
g. an abnormality of condition that has been present from birth
h. chest pain caused by the restriction of oxygen; strangling, squeezing, spasm
i. high potassium level
j. blood clot
k. blood pressure that is consistently higher than normal
l. moving in a reverse direction, opposite of the normal pathway
m. generated by self, self generated
n. event that can be seen to both start and stop, happening in short episodes
o. pacing function designed to give heart ample opportunity to provide own intrinsic stimulation
p. the lowest setting a pacemaker pulse generator will pace
q. the relaxation and refilling phase of the heart after the depolarization period
r. pacemaker mode that senses to activity level of an individual and adjusts the rate automatically
s. an occurrence where the pacing impulse and intrinsic beat happen simultaneously
t. to remove by surgery, melt away, vaporize, wear away
u. the transmission process of an electrical impulse
v. the penetration & entering of the body such as surgery
w. tissue damage caused by lack of oxygen rich blood
x. a stiffening / enlargement (dilation) of the heart muscle
y. heart rhythm precipitated above the ventricles
z. protrusion in a compromised wall section of a blood vessel

60. This condition can be caused by inflammation of the pericardium (thin sack of fluid that surrounds the heart) that puts constrictive pressure on the hearts surface blood vessels and limits its ability to expand during the repolarization process
    a. edema
    b. tamponade
    c. constrictive infection
    d. dyspnea

61. Chronic Obstructive Pulmonary Disease can be associated to the following diseases
    a. chronic bronchitis
    b. asthma
    c. emphysema
    d. all of the above

62. Edema is the result of the fluid portion of oxygen poor blood in the veins being forced into the surrounding tissue making it swell. This condition can be caused due to the following reasons (circle all that apply)
    a. kidney failure
    b. liver problems
    c. sinus bradycardia
    d. heart disease
    e. transient ischemic attack (TIA)
    f. all of the above

63. Select the proper answers to match with the right definition of a particular form of edema
    a. anasarca b. mechanical c. pulmonary d. dependant e. Lymphedema f. pitting
    _ general swelling throughout the body
    _ caused by tight fitting clothing such as socks, underwear, etc.
    _ this type can be experienced at the end of the day after being in a sitting position for long periods of time
    _ will leave a depression in the swollen tissue after pressing down on it with finger tips
    _ is caused by blockages in the body's drainage channels
    _ involves the respiratory system re. build-up of fluid in the lungs

64. AV heart block is a situation in which the electrical impulse conduction between the atria to the ventricles has been delayed or blocked, resulting in an arrhythmia that produces missed complexes and a potential lack of synchronization of the hearts mechanical activity. Some possible causes of this condition can include but are not limited to (may be more than one answer)
    a. medications ex. digitalis, calcium channel blockers, beta blockers
    b. diseased tissue in the AV node
    c. prior heart attack
    d electrolyte imbalance
    e. all of the above

65. The walls of the left ventricle are ___ times as thick as the right ventricle to provide a stronger catalyst to move the blood throughout the body
    a. two
    b. four
    c. three
    d. the same thickness
    e. none of the above

66. The mitral valve located between the left atria and left ventricle has 4 flaps
    a. true
    b. false

67. The right atria sends blood to the right ventricle through the tricuspid valve which has two flaps
    a. true
    b. false

Provide an interpretation of the following rhythm descriptions without the use of a rhythm strip tracing using the established guidelines only

68. PRI .20 QRS .10 QT .48 R-R .80 HR @ 70 rhythm is regular with upright and rounded P-waves @ one per QRS, T-waves are upright and rounded

69. PRI .18 QRS .13 QT .40 R-R .85 HR @ 66 rhythm is regular with upright and rounded P-waves @ one per QRS, T-waves are upright and rounded, premature atrial beats that occur every third cardiac cycle

70. PRI variable QRS .14 QT.42 R-R .84 HR @ 69 rhythm is regular with P-waves that display multiple different appearances @ one per QRS, T-waves are upright and rounded, premature ventricular beats that possess the same appearance but have episodes involving group beating in runs of six

71. Irregular rhythm with a non-recognizable P-wave QRS .15 QT .40 R-R variable, erratic isoelectric baseline, HR @ 110 with episodes with HR up to 150s

72. Irregular rhythm with a saw-tooth isoelectric baseline appearance, QRS .08 R-R-variable, HR at 68, premature ventricular beats that display different Appearances, occurring every other cardiac cycle

73. PRI .11 QRS .10 QT .40 R-R . 82 HR @ 68 rhythm is regular with inverted P-waves @ one per QRS, T-waves are upright and rounded

74. PRI .20 QRS .10 QT .42 R-R .84 HR @ 56 bpm w/ intermittent drops down to 42 bpm constant PRI, ventricular rhythm may be irregular with upright and rounded P-waves that vary from one to runs with two per QRS, T- waves are upright and rounded

75. PRI variable QRS .10 QT.35 R-R variable HR @ 115 heart rhythm is irregular with P-waves that display different appearances, T-waves are upright and rounded, this rhythm also possesses varying PRI measurements, early ventricular complexes that have different appearances that occur every fourth cardiac cycle and have episodes of group beating in sets of two and three

76. PRI variable QRS .10 QT .50 R-R 1.00 HR @ 38 both atrial and ventricular complexes march out, but have no synchronization to each other, the P-waves are upright and rounded and count out to be variable in number to the QRS complexes (ex. 2:1 3:1), there are episodes when the ventricular HR drops into the 20s with R-R measuring out to be of multiple second time durations (please explain how/why you arrived at the conclusions you came to)

77. This arrhythmia inserts itself in and out of an underlying rhythm without staying for a sustained period of time. In this rhythm there is nothing measurable with the ventricular complexes appearing to spiral around the isoelectric baseline with changes in **vertical height** from wide to narrow signifying amplitude changes, the HR is @ 185 bpm

78. Which of the following arrhythmias would be classified to be in the category of supraventricular tachycardia (circle the correct answers)
a. paroxysmal atrial tachycardia
b. uncontrolled atrial fibrillation
c. junctional tachycardia
d. ventricular tachycardia
e. torsades de pointes
f. multi-focal atrial tachycardia
g. atrial flutter

79. Idioventricular rhythm possesses the following characteristic(s) (may be more than one answer)
a. is also referred to as ventricular tachycardia with a heart rate below 120 bpm
b. having a complete and total failure of all pacemaker sites above the ventricles
c. possesses a heart rate in the range of 20-40 bpm
d. has a T-wave with a deflection in the opposite direction of the QRS

80. In Ventricular Tachycardia the impulses of the sinoatrial node are over-ridden by abnormal electrical impulses formed within the ventricles
a. true
b. false

81. The minimum time duration that Ventricular Tachycardia is considered to be a sustained arrhythmia is
a. one minute
b. fifteen seconds
c. thirty seconds
d. ninety seconds

82. Select from the following, any potential problem(s) that may be induced by the arrhythmia Ventricular Tachycardia
(may be more than one answer)
a. diminished blood supply to the body
b. progression to cardiac arrest
c. progression to ventricular fibrillation
d. fatality

83. The pacemaker is designed to manage the conduction system in situations when the hearts intrinsic impulses do not sustain an adequate number of beats per minute (extreme bradycardia).
    a. true
    b. false

84. Three seconds as measured on a rhythm strip would include this number of larger boxes
    a. 20
    b. 30
    c. 15
    d. 10

85. Asystole is a situation where there is no electrical activity present to stimulate a myocardial response creating no blood perfusion to the body. The only intervention to this condition is the utilization of electrical shock
    (explain why this conclusion was arrived at)
    a. true
    b. false

86. The heart is designed to produce simultaneous contractions of both ventricles together. In conditions such as Congestive Heart Failure, heart disease, or an aged conduction system, this synchronized action is affected. Explain why this lack of synchronization would affect the body. (Provide 5 possible side effects)
    1)
    2)
    3)
    4)
    5)

87. Treatment for this conduction abnormality mentioned above would include
    a. the implementation of an implantable cardioverter defibrillator
    b. the implementation of a dual chamber pacemaker
    c. the implementation of a bi-ventricular pacemaker
    d. all of the above

88. The halter heart monitor is a vest that worn by an individual to record cardiac activity for a time period of twenty-four hours or more while the patient performs their normal daily activities to learn what information **a**. pacemaker function **b**. possible reasons for chest pain **c**. effects of new cardiac medications **d**. provide a record of occurring arrhythmias **e**. monitor post heart attack patients
    a. a, b, d
    b. c & d only
    c. b, c, d, e
    d. all of the above

89. To obtain an immediate and precise tracing of abnormal cardiac electrical activity occurring with a patient, the following test is performed
    a. twelve lead electrocardiogram
    b. echocardiogram
    c. catherization
    d. telemetry tracing

90. A premature ventricular complex that does not affect the timing of a rhythm by allowing enough time to permit the next intrinsic beat fall into the right timing is called a
    a. multi-focal premature ventricular complex
    b. uni-focal premature ventricular complex
    c. premature ventricular complex with a compensatory pause
    d. ventricular escape beat

91. Arterial blood gases (ABGs) are done from an artery to examine gas levels present within on blood that is in its freshly re-oxygenated state
    a. true
    b. false

92. The main concern for adverse problems related to the heart condition known as Long QT Syndrome would be
    a. the increase in fluid build-up in the body's tissue
    b. early ectopic beats falling into the previous ventricular repolarization cycle, inducing a fatal rhythm
    c. increased atrial activity with an increased heart rate

93. In group beating of Premature Ventricular Complexes, more than _____ in a run is considered Ventricular Tachycardia
    a. three
    b. four
    c. five
    d. six

94. Atrio-Ventricular Dissociation displays which characteristic on a cardiac monitor
    a. both atrial and ventricular rhythms are regular but are not synchronized with each other
    b. there is an atrial depolarization missing with every third cardiac cycle
    c. there are pacing impulse spikes that are not associated with the atria or ventricular depolarizations
    d. the pacing impulse spikes are not present with the heart rate below the standby rate

95. Sinus Tachycardia is described as having a heart rate that exceeds the normal rate of one hundred beats per minute. Some of the possible reasons for this occurrence may include but are not limited to
(may be more than one answer)
a. excessive use of nicotine and caffeine
b. alcohol consumption
c. body's response to fever, pain, or dehydration
d. vagal response
e. all of the above

96. This arrhythmia has a transient episode that involves a series of four P-waves without QRS's following them, with the PRI intervals before and after the occurrence regular without lengthening or lessening. This would be described as
a. sinus pause
b. sinus arrest
c. no ventricular response
d. sinoatrial exit block

97. The correct rate for accelerated idioventricular rhythm is
a. 20-40 bpm
b. 40-60 bpm
c. 40-100 bpm
d. 100-120 bpm

98. In first degree AV block the guidelines are as follows
a) .21-.24 slight b) .25-.29 moderate c) greater than .30 considered severe
a. true
b. false

99. In 2nd Degree Type 1 Sinoatrial Heart Block , the PR interval lessens instead of increasing as seen in 2nd Degree Type 1 Atrioventricular Heart Block's progressive lengthening PR interval
a. true
b. false

100. The hearts conduction system begins with an impulse generated at the site of the _ where it travels through the _ to initiate atrial contractions, it then follows the _ that brings it to the _, through the _, down the _, to the _ where it stimulates a ventricular contraction
a. purkinje fibers
b. right and left bundle branches
c. atrioventricular node
d. intra-atrial pathway
e. sinoatrial node
f. intra-nodal pathway
g. bundle of his

103. The difference(s) between sinus pause/sinus arrest and slow ventricular response is / are
   a. there is no electrical activity present in the atria in sinus pause, where with as slow ventricular response there is
   b. sinus pause can occur in sinus rhythms only
   c. in sinus pause the problem lies in the sinoatrial node, with slow ventricular response the problem lies in most instances within the AV node conducting the impulse to the ventricles
   d. a & c
   e. all of the above

104. The procedure utilized to repair the adjacent conduction pathway that causes conduction problems and precipitates arrhythmias in the condition known as Wolff Parkinson White Syndrome is referred to as
   a. balloon angioplasty
   b. cardioversion
   c. radiofrequency ablation
   d. revised conduction therapy

NOTES

104 Question
Answer Key

Answer Key

1. b
2. b
3. a
4. a
5. b
6. a
7. a
8. c
9. b
10. a
11. c
12. a
13. b *
14. a *
15. b
16. a
17. b
18. d
19. c
20. b
21. b *
22. b
23. c *
24. c
25. a
26. a
27. b
28. c
29. a
30. a
31. e
32. a
33. b
34. a
35. b
36. a *
37. a *
38. b, c *
39. c
40. a *
41. b
42. b
43. e
44. a
45. c
46. a
47. b
48. a,e,b,c,d
49. h *
50. c
51. part 1 a *
part 2 c
52. b *
53. b, c *
54. c
55. b *
56. b *
57. b, d *
58. a, b *
60. b
61. d
62. a, b, d
63. a,b,d,f,e,c
64. e
65. c *
66. b *
67. b *
78. a, b, c, f *
79. b, c, d
80. a
81. c
82. a, b, c, d (all) *
83. a *
84. c
85. b *
87. c
88. d
89. a *
90. c
91. a
92. b
93. a
94. a
95. a, b, c
96. c *
97. c
98. a
99. a *
100. e, d, f, c, g, b, a
101. e,c,f,b,g,h,a,i,j,d,b,l,k
102. c
103. e
104. c

59. f, m, q, n, w, I, z, g, o, s, y, a, e, d, l, b, p, c, r, t, u, v, j, x, h, k
68. sinus rhythm @ 70 w/ prolonged QT interval
69. sinus rhythm @ 66 w/ bundle branch block, PAC's w/ atrial tri-geminy
70. wandering atrial pacemaker @ 69 w/ uni-focal PVC's w/ six beat runs of v-tach
71. uncontrolled a-fib @ 110 w/ bundle branch block, episodes of rapid a-fib w/ HR ^ 150's
72. controlled variable conduction a-flutter @ 68 w/ multi-focal PVC's w/ ventricular bi-geminy
73. accelerated junctional @ 68
74. second degree AV block type ll @ 56
75. multi-focal atrial tachycardia @ 115 w/ PVC's w/ ventricular bi-geminy, couplets, triplets
76. Third Degree AV block @ 38
77. Torsades de Pointes @ 185
86. the lack of synchronization between the contractions of the right and left ventricles diminishes the force applied to the perfusion by making the left ventricle walls contract at different times. This in turn provides less blood supply to the body and its organs, making the heart work harder to make up the difference, leading to eventual heart failure.

Bold Numbers w/asterisk required further explanations that can be found on the reverse side

13. They are measured to determine time durations between ventricular depolarizations.
14. More specifically as interferrance at the site of the AV node. All impulses do eventually make it to the ventricles. This is considered more of a conduction delay.
21. Associated with a sinoatrial node abnormality. A-Flutter and A-Fib are not sinus rhythms. A time duration more than 3 seconds is considered sinus arrest.
22. PVC's can occur in all rhythms. This form of ectopy can be the result of irritated ventricles, lack of oxygen rich blood supply to the heart, or external factors such as stress, medications, anxiety, or caffeine
23. The junctional conduction is retrograde, causing the atria to be conducted by a reverse impulse. This makes the P-wave inverted.
36. With a total block of conduction between the atria and ventricles, specialized cells within in the ventricles create their own impulse as an escape mechanism to keep blood moving throughout the body. This causes them to lose all synchronization with the atria.
37. Sinus references to the site of the disturbance, the sino-atrial node. Diseased tissue inhibits regular and rhythmatic creation of the electrical impulses.
38. Mostly seen as a transient arrhythmia entering and leaving a base rhythm that does not sustain for long periods of time.
40. Disorganized electrical activity makes the ventricles quiver instead of their normal contractions. It is estimated that an individual cannot last more than four minutes in this arrhythmia without serious organ damage or death. Electrical shock provides an interruption in the electrical activity to provide an opportunity for the hearts natural pacemaker (SA node) to reassume control of the hearts conduction system.
48. AV & VA delays refer to the set time durations programmed into the pacemaker.
49. This condition is hard to pre-diagnose (without the aid of an elective EKG (Delta Wave observation), due to the fact that it doesn't display symptoms when not actively causing a disturbance. This abnormality is diagnosed mainly as an incident occurs.
52. Slower rates can be seen in situations involving salvos of PVC's.
53. Early ectopic beats that occur in the cells of the ventricular repolarization phase of the cardiac cycle can initiate an response contrary to that of a normal repolarization.
55. The electric shock is applied at the proper timing to provide an interruption in the disorganized intrinsic electrical stimulation causing the arrhythmia, giving the sinoatrial node the opportunity to reset & re-establish itself as the pacemaker.
56. Referred to as sinus pause, a flutter and a fib are not considered sinus rhythms.
58. Bi-ventricular pacemaker stimulates simultaneous contractions with both ventricles lessening the workload on the heart. Came about in 2002. Also referred to as "Cardiac Resynchronization Therapy."
65. The thicker walls provide more strength for perfusion due to their firmness which provides for less flexing during depolarization.
66. The mitral valve has two flaps.
67. The tricuspid valve has three flaps.
78. Ventricular Tachycardia and Torsades de Pointes are governed by the ventricles, controlled A-Flutter possesses a heart rate that is below 100 beats per minute.
82. Due to the rapid depolarization of the ventricles there is no chamber refill time, providing less blood supply to the body.

83. The pacemaker is designed to manage patients with low heart rates (extreme bradycardic). The defibrillator is designed to break rapid heart rates commonly precipitated by chronic uncontrolled cardiac arrhythmias such as Atrial Fibrillation
85. The heart is in a depolarized state in asystole, shocking would not be beneficial. The ability for the sinoatrial node to create electrical impulse must be present to utilize electrical shock.
86. the lack of synchronization between the contractions of the right and left ventricles diminishes the force applied to the perfusion by making the left ventricle walls contract at different times. This weakened perfusion in turn provides less blood supply to the body and its organs, making the heart work harder to compensate for the difference, leading to eventual heart failure.
89. Telemetry provides only a limited view, but can lead to the order of a twelve lead EKG and further investigate changes. Physicians rely on the multiple different views of the twelve lead as being more precise.
96. the regularity of the PR intervals in this example rule out wenckebach, which has a progressively lengthening PR interval.

Notes

Notes

Index

P

R

S

T

U

V

W

"Take advantage of two more opportunities offered by this author to learn more about heart related diseases, conditions, treatments and procedures!"

"Cardiac Telemetry Basics"
ISBN # 9780615151526
140 pages

Written and formatted to assist the new healthcare professional to integrate into the cardiac field. This first volume in the series of three provides its reader with the characteristics, guidelines and information necessary to learn the interpretation of heart rhythms. It also includes heart physiology and related material that will help one to understand what induces the heart to alter its behavior. This book which is available and is sold worldwide has helped to present telemetry in a format that comes complete with actual rhythm strip tracings that are utilized to help describe what is being seen and explained as it is observed on the heart monitor. This is an excellent informational source for those wanting to learn cardiac healthcare as well as those professionals who might want to explore other possible healthcare avenues or just expand their knowledge. As with other books in the series this volume includes sections that include cardiac diseases and conditions, testing and diagnostic methods, corrective procedures, terminology and a question and answer section that follows at the end to provide a means for you to perform a self-evaluation of your comprehension of the reviewed material.

"Cardiac Telemetry Basics Test Book"
ISBN # 9781435754744
56 pages

This book was designed and written to provide the means for the individual to perform a self-evaluation of the material that has been previously reviewed in the book "Cardiac Telemetry Basics." Designed in an easy non-formal format, this book is comprised mainly of multiple choice questions (with a few mind twisters and trick questions) to help to enhance and complement the previously learned material. Complete with an answer key and author comment section, this is an excellent teaching, interactive and learning tool that will show particular areas that are in need of more study. It is a must have book to accompany "Cardiac Telemetry Basics."

These books are also available at: http://stores.lulu.com/ohkwaliliterature
or @ ohkwaliliterature@verizon.net

Send comments to: ohkwaliliterature@verizon.net

Order Form

Like this book? Order a copy for a friend below!

* Shipping Information (please write clearly)
  * allow 2 – 3 weeks for delivery

Name ____

Address ____

City ___________________________ State_______________

Zip Code _________ - _________

* Payment Information

( ) Personal Check (must clear before shipping) ( ) Money Order ( ) Pay Pal

A. Cardiac Telemetry Basics
Order Quantity ____ x $21.95 = ______ Sub Total ______

B. Cardiac Telemetry Self-Evaluation Comprehension Test Book
Order Quantity ____ x $13.95 = ______ Sub Total ______

C. Heart Notes
Order Quantity ____ x $ 24.95 = ______ Sub Total ______

Shipping: $4.95 for first book, please add $2.00 for each additional book (bulk rates available per quantity orders) S&H ______

New York State residents please add appropriate sales tax *Tax* ______

Total ________

Please send order form along with payment to: **Ohkwali' Literature**
710 Herkimer Road
Utica, New York 13502

or place your order at:
http://stores.lulu.com/ohkwaliliterature
or
ohkwaliliterature@verizon.net